T0155633

METHODS IN MOLECULAR BIOLOGY

Series Editor
John M. Walker
School of Life and Medical Sciences
University of Hertfordshire
Hatfield, Hertfordshire, AL10 9AB, UK

For further volumes:
http://www.springer.com/series/7651

Mitophagy

Methods and Protocols

Edited by

Nobutaka Hattori and Shinji Saiki

Department of Neurology, Juntendo University School of Medicine, Hongo, Bunkyo, Tokyo, Japan

Editors
Nobutaka Hattori
Department of Neurology
Juntendo University School
of Medicine
Hongo, Bunkyo, Tokyo, Japan

Shinji Saiki
Department of Neurology
Juntendo University School
of Medicine
Hongo, Bunkyo, Tokyo, Japan

ISSN 1064-3745 ISSN 1940-6029 (electronic)
Methods in Molecular Biology
ISBN 978-1-4939-9274-4 ISBN 978-1-4939-7750-5 (eBook)
https://doi.org/10.1007/978-1-4939-7750-5

Printed on acid-free paper

This Humana Press imprint is published by the registered company Springer Science+Business Media, LLC part of Springer Nature.
The registered company address is: 233 Spring Street, New York, NY 10013, U.S.A.

Preface

In 1955, Christian de Duve discovered previously unknown organelles: lysosomes. The discovery significantly contributed to him being awarded a share of the 1974 Nobel Prize in Physiology or Medicine for elucidating "the structural and functional organization of the cell." More recently, Yoshinori Ohsumi has won the 2016 Nobel Prize in Physiology or Medicine for discovering and elucidating mechanisms underlying autophagy, a fundamental process for degrading and recycling cellular components. Thus, this volume for the methods in molecular biology which focuses on mitophagy, a form of autophagy, is timely and deeply meaningful for the readers. In other words, mitophagy is autophagy of mitochondria.

While eukaryotic cells have acquired the highly efficient power-generating system of aerobic respiration by incorporating mitochondria into the cytosol, alleviation of oxidative stress by appropriate regulation of mitochondria has become an emerging and inevitable issue. Non-dividing cells or tissues with high-energy demands in long-lived animals such as humans have particular a need to regulate mitochondrial activity to avoid the production of deleterious reactive oxygen species. Mitochondrial dysregulation is now implicated in various human diseases including cancer, diabetes, myopathy, and neurodegeneration such as Parkinson's disease and Alzheimer's disease. Cellular damage occurs in response to genetic perturbations, nutrient deprivation, aging, and environmental toxins. The task of managing general and specific cellular damage is largely under the control of the highly regulated process called autophagy. The term autophagy is used to describe lysosomal-mediated degradation of intracellular contents, which can be divided into three basic mechanisms: (1) chaperone-mediated autophagy, (2) microautophagy, and (3) macroautophagy. Firstly, chaperone-mediated autophagy, initiated by chaperone Hsc70, recognizes one protein at a time, and Hsc70 carries the protein to the lysosomes via binding to the lysosomal-associated membrane protein. Secondly, microautophagy involves invagination of lysosomal membranes to encircle cellular contents that may include proteins and lipids. Lipids, proteins, or organelles can be degraded through this pathway. Whether lipids, organelles, and other proteins are marked by specific modifications to be recognized by the lysosomes is highly likely; however, the majority of these have yet to be defined. Thirdly, macroautophagy is the most extensively studied form of autophagy, which involves formation of double-membrane structures that encircle proteins, lipids, and organelles. Degradation of mitochondria through the macroautophagy pathway is also termed "mitophagy." In addition, degradation of other cellular structures, such as fragments of the nucleus, lipid droplets, peroxisomes, ribosomes, and endoplasmic reticulum, has also been called nucleophagy, lipophagy, pexophagy, ribophagy, and reticulophagy.

Mitochondrial dysregulation has indeed been implicated in various human diseases, including cancer, diabetes, myopathy, and a variety of neurodegenerative disorders such as amyotrophic lateral sclerosis, Huntington's disease, and Parkinson's disease. Therefore, mitochondrial maintenance involves mitophagy, a selective autophagy process that removes abnormal mitochondria.

Finally, the chapters in this volume are concerned not just with methodology, but also provide the background for the readers to appreciate the importance of monitoring mitophagy with regard to the particular questions being asked. This volume consists of 15 chapters that are focused on the analysis of mitophagy in connection with Parkinson's disease, especially young-onset Parkinson's disease impaired by PINK/parkin pathway, Drosophila, yeast, and *Caenorhabditis elegans* models.

One hope is that this volume will stimulate and encourage researchers to carry out ongoing and new investigations into mitophagy so that we continue to strive to understand the exact mechanisms of these complex systems.

Contents

Part III Others

Contributors

WADO AKAMATSU • *Center for Genomic and Regenerative Medicine, Juntendo University School of Medicine, Tokyo, Japan*

SATOKO ARAKAWA • *Department of Pathological Cell Biology, Medical Research Institute, Tokyo Medical and Dental University, Tokyo, Japan*

FOLMA BUSS • *Cambridge Institute for Medical Research, University of Cambridge, Cambridge, UK*

NIKOLAOS CHARMPILAS • *Institute of Molecular Biology and Biotechnology, Foundation for Research and Technology-Hellas, Heraklion, Crete, Greece; Department of Biology, University of Crete, Heraklion, Crete, Greece*

HAO CHEN • *Institute of Neurology, Guangdong Key Laboratory of Age-related Cardiac-cerebral Vascular Disease; The Affiliated Hospital of Guangdong Medical University, Guangdong Medical University, Zhanjiang, China; Medical Research Center, The First Affiliated Hospital of Zhengzhou University, Zhengzhou, China*

IVAN DIKIC • *School of Medicine, University of Split, Split, Croatia; Institute of Biochemistry II, Goethe University Medical School, Frankfurt, Germany; Buchmann Institute for Molecular Life Sciences, Frankfurt, Germany*

DU FENG • *Institute of Neurology, Guangdong Key Laboratory of Age-related Cardiac-cerebral Vascular Disease, The Affiliated Hospital of Guangdong Medical University, Guangdong Medical University, Zhanjiang, China; Medical Research Center, The First Affiliated Hospital of Zhengzhou University, Zhengzhou, China; The Department of Developmental Biology, Harvard School of Dental Medicine, Harvard Medical School, Boston, MA, USA*

MOTOKI FUJIMAKI • *Department of Neurology, Juntendo University School of Medicine, Tokyo, Japan*

KENTARO FURUKAWA • *Department of Cellular Physiology, Niigata University Graduate School of Medical and Dental Sciences, Niigata, Japan*

NORIHIKO FURUYA • *Department of Neurology, Juntendo University School of Medicine, Hongo, Bunkyo, Tokyo, Japan*

NOBUTAKA HATTORI • *Department of Neurology, Juntendo University School of Medicine, Hongo, Bunkyo, Tokyo, Japan*

SHINYA HONDA • *Department of Pathological Cell Biology, Medical Research Institute, Tokyo Medical and Dental University, Tokyo, Japan*

ATSUSHI HOSHINO • *Department of Cardiovascular Medicine, Graduate School of Medical Science, Kyoto Prefectural University of Medicine, Kamigyo-ku, Kyoto, Japan*

YUZURU IMAI • *Department of Research for Parkinson's Disease, Juntendo University Graduate School of Medicine, Tokyo, Japan; Department of Neurology, Juntendo University Graduate School of Medicine, Tokyo, Japan*

YOKO IMAMICHI • *Department of Neurology, Juntendo University School of Medicine, Tokyo, Japan*

TSUYOSHI INOSHITA • *Department of Treatment and Research in Multiple Sclerosis and Neuro-intractable Disease, Juntendo University Graduate School of Medicine, Tokyo, Japan*

KEI-ICHI ISHIKAWA • *Department of Neurology, Juntendo University School of Medicine, Tokyo, Japan*

TOMOTAKE KANKI • *Department of Cellular Physiology, Niigata University Graduate School of Medical and Dental Sciences, Niigata, Japan*
CHIEKO KISHI-ITAKURA • *Cambridge Institute for Medical Research, University of Cambridge, Cambridge, UK*
DANIEL J. KLIONSKY • *Department of Molecular, Cellular and Developmental Biology, Life Sciences Institute, University of Michigan, Ann Arbor, MI, USA*
KONSTANTINOS KOUNAKIS • *Institute of Molecular Biology and Biotechnology, Foundation for Research and Technology-Hellas, Heraklion, Crete, Greece; Department of Biology, University of Crete, Heraklion, Crete, Greece*
WEN LI • *Institute of Neurology, Guangdong Key Laboratory of Age-related Cardiac-cerebral Vascular Disease, The Affiliated Hospital of Guangdong Medical University, Guangdong Medical University, Zhanjiang, China; Medical Research Center, The First Affiliated Hospital of Zhengzhou University, Zhengzhou, China*
SHUPENG LI • *Institute of Neurology, Guangdong Key Laboratory of Age-related Cardiac-cerebral Vascular Disease, The Affiliated Hospital of Guangdong Medical University, Guangdong Medical University, Zhanjiang, China; Medical Research Center, The First Affiliated Hospital of Zhengzhou University, Zhengzhou, China*
GUANGHONG LIN • *Institute of Neurology, Guangdong Key Laboratory of Age-related Cardiac-cerebral Vascular Disease, The Affiliated Hospital of Guangdong Medical University, Guangdong Medical University, Zhanjiang, China; Medical Research Center, The First Affiliated Hospital of Zhengzhou University, Zhengzhou, China*
XU LIU • *Department of Molecular, Cellular and Developmental Biology, Life Sciences Institute, University of Michigan, Ann Arbor, MI, USA*
SATOAKI MATOBA • *Department of Cardiovascular Medicine, Graduate School of Medical Science, Kyoto Prefectural University of Medicine, Kamigyo-ku, Kyoto, Japan*
HONGRUI MENG • *Research Institute for Diseases of Old Age, Juntendo University Graduate School of Medicine, Tokyo, Japan*
MINORU NAGI • *Department of Chemotherapy and Mycoses, National Institute of Infectious Diseases, Tokyo, Japan*
SACHIYO NAGUMO • *Graduate School of Frontier Biosciences, Osaka University, Osaka, Japan*
IVANA NOVAK • *School of Medicine, University of Split, Split, Croatia*
KOJI OKAMOTO • *Graduate School of Frontier Biosciences, Osaka University, Osaka, Japan*
HIDEYUKI OKANO • *Department of Physiology, Keio University School of Medicine, Shinjuku, Tokyo, Japan*
SHINJI SAIKI • *Department of Neurology, Juntendo University School of Medicine, Hongo, Bunkyo, Tokyo, Japan*
YUKIKO SASAZAWA • *Department of Neurology, Juntendo University School of Medicine, Tokyo, Japan*
SHIGETO SATO • *Department of Neurology, Juntendo University Graduate School of Medicine, Hongo, Bunkyo, Tokyo, Japan*
MIYUKI SATO • *Laboratory of Molecular Membrane Biology, Institute for Molecular and Cellular Regulation, Gunma University, Maebashi, Gunma, Japan*
KEN SATO • *Laboratory of Molecular Traffic, Institute for Molecular and Cellular Regulation, Gunma University, Maebashi, Gunma, Japan*
KAHORI SHIBA-FUKUSHIMA • *Department of Treatment and Research in Multiple Sclerosis and Neuro-intractable Disease, Juntendo University Graduate School of Medicine, Tokyo, Japan*

SHIGEOMI SHIMIZU • *Department of Pathological Cell Biology, Medical Research Institute, Tokyo Medical and Dental University, Tokyo, Japan*
MATILDA ŠPRUNG • *Faculty of Science, University of Split, Split, Croatia*
KATSUHIKO SUMIYOSHI • *Department of Nutrition, Tokiwa University, Mito, Ibaraki, Japan*
KOICHI TANABE • *Department of Food Science and Human Nutrition, Faculty of Agriculture, Ryukoku University, Otsu, Shiga, Japan; Department of Chemotherapy and Mycoses, National Institute of Infectious Diseases, Tokyo, Japan*
NEKTARIOS TAVERNARAKIS • *Institute of Molecular Biology and Biotechnology, Foundation for Research and Technology-Hellas, Heraklion, Crete, Greece; Department of Basic Sciences, Faculty of Medicine, University of Crete, Heraklion, Crete, Greece*
SATORU TORII • *Department of Pathological Cell Biology, Medical Research Institute, Tokyo Medical and Dental University, Tokyo, Japan*
MASATSUNE TSUJIOKA • *Department of Pathological Cell Biology, Medical Research Institute, Tokyo Medical and Dental University, Tokyo, Japan*
AKIHIRO YAMAGUCHI • *Center for Genomic and Regenerative Medicine, Juntendo University School of Medicine, Tokyo, Japan*
SHUN-ICHI YAMASHITA • *Department of Cellular Physiology, Niigata University Graduate School of Medical and Dental Sciences, Niigata, Japan*
ZHIYUAN YAO • *Department of Molecular, Cellular and Developmental Biology, Life Sciences Institute, University of Michigan, Ann Arbor, MI, USA*

Part I

PINK 1/Parkin-Mediated Mitophagy

Methods in Molecular Biology (2018) 1759: 3–8
DOI 10.1007/7651_2017_38

Published online: 12 May 2017

Short Overview

Norihiko Furuya

Abstract

Mitochondrial autophagy (mitophagy) is a mitochondrial quality control mechanism that selectively removes damaged mitochondria via autophagic degradation. Autophagic adaptor/receptor proteins contribute to the selective degradation of damaged mitochondria by autophagy. A part of them containing both ubiquitin binding domains and Atg8 interacting motif (AIM)/LC3 interacting region (LIR) motifs, which bind to the autophagy-related protein 8 (Atg8) family (LC3 and GABARAP family), lead ubiquitylated (damaged) mitochondria to selective removal. On the other hand, some specific outer mitochondrial membrane-anchored proteins containing AIM/LIR motif function as another type of autophagy adaptor/receptor proteins. Here I briefly summarize mechanisms of mitophagy and its related proteins.

Keywords Atg8 interacting motif (AIM)/LC3 interacting region (LIR), Autophagy adaptor, Mitophagy

Mitochondrial dysfunction is tightly related with aging, cancer, diabetes, and neurodegenerative disease including Parkinson's disease [1–5]. Mitochondrial quality is strictly controlled by multiple mechanisms to maintain homeostasis and functions of mitochondria. Mitochondrial autophagy (mitophagy) is a mitochondrial quality control mechanism that selectively encloses damaged mitochondria within autophagosomes and delivers them to lysosomes for degradation.

Autophagy is an evolutionarily conserved membrane dynamic process contributing to the degradation of intracellular molecules and organelle [6–8]. Although autophagy is strongly induced under nutrient-poor environments to supply amino acids for cell survival, autophagy also participates in the selective clearance of misfolded or aggregated proteins, damaged organelles, and invaded pathogens [6, 8–10]. Those selective autophagic cargos are labeled with polyubiquitin chain. A part of autophagy adaptor/receptor proteins such as p62/SQSTM-1, NBR1, OPTN, NDP52, and TAX1BP1 containing both ubiquitin binding domains and LC3 interacting region (LIR) motifs, which bind to the autophagy-related protein 8 (Atg8) family (LC3 and GABARAP family), contribute to recognition of ubiquitylated cargo in selective autophagy [11]. On the other hand, some specific outer mitochondrial membrane-anchored proteins such as Atg32, Bnip3, Nix/Bnip3L, and FUNDC1 function as mitophagy adaptor/receptor proteins. They mediate selective clearance of damaged or superfluous mitochondria.

PINK1/Parkin-mediated mitophagy is one of the best-characterized mechanisms of ubiquitin-dependent mitophagy in mammalian cells [12]. A mitochondrial Ser/Thr kinase PINK1 and a ubiquitin E3 ligase Parkin are commonly mutated in autosomal recessive juvenile Parkinson's disease, and the defects of PINK1/Parkin-mediated mitophagy are thought to be a cause of a part of Parkinson's disease onsets. In dysfunctional mitochondria, which lose the membrane potential across the inner mitochondrial membrane (IMM) produced by oxidative-phosphorylation complexes, the protease activity of presenilin-associated rhomboid-like (PARL), which mediates cleavage of PINK1, is reduced, and subsequently PINK1 is stabilized on the outer mitochondrial membrane (OMM) associating with TOMM complex [13, 14]. Stabilized PINK1 phosphorylates Parkin on Ser65 in N-terminal ubiquitin-like (UBL) domain, and conserved Ser65 of ubiquitin on OMM [15–18]. Phosphorylation of Parkin activates its E3 ubiquitin ligase activity and induces recruitment of Parkin to damaged mitochondria. Phosphorylated ubiquitin also activates Parkin E3 ligase activity. Various OMM proteins are ubiquitylated by Parkin [19–21]. The feed-forward loop of phospho-ubiquitin chains generation on damaged mitochondria, a notable feature of PINK1/Parkin-mediated mitophagy, dramatically enhances subsequent autophagic engulfment of damaged mitochondria. Following ubiquitylation of OMM proteins by Parkin, a part of OMM proteins, such as MFN1, MFN2, and MitoNEET, are degraded by proteasome [20, 21]. On the other hand, IMM and matrix proteins are degraded in an autophagy-dependent manner [21]. Autophagy adaptor/receptor proteins such as p62/SQSTM-1, NBR1, OPTN, NDP52, and TAX1BP1 contribute to recognition of ubiquitylated mitochondria in mitophagy. Although all of them are not essential for PINK1/Parkin-mediated mitophagy, not only the primary adaptors such as OPTN and NDP52 (TAX1BP1 to some extent) for PINK1/Parkin-mediated mitophagy, but nonessential adaptors such as p62 and NBR1 also contribute to the efficiency of autophagic engulfment of damaged mitochondria [22, 23]. In addition, either OPTN or NDP52, but not p62 recruit upstream autophagy proteins (i.e., ULK1, DFCP1) to mitochondria PINK1 (or phospho-ubiquitin)-dependently and Parkin-independently [22]. Taken together, PINK1/Parkin system functions as an enhancer for the selectivity of damaged mitochondria by autophagic machinery, and may have important roles in tissues easily exposed by mitochondrial stresses (i.e., dopaminergic neurons).

When budding yeasts are cultured in non-fermentable medium for prolonged period, mitophagy is induced. Mitophagy in yeast depends on Atg32, a single pass OMM protein [24, 25]. Atg32 is strongly induced and accumulates on the surface of mitochondria during respiratory growth. Cytosolic domain of Atg32 contains an Atg8-family interacting motif (AIM, equivalent to mammalian LIR) for binding to Atg8. Atg32 also has a consensus motif important for binding to Atg11, a scaffold protein for selective autophagy in yeast. Atg32-Atg11 interaction on the surface of mitochondria is an initial event, and recruits core Atg proteins as a platform for assembly to form mitophagosome.

As is the case of Atg32 in yeast, there are some mitophagy receptors localized at the outer membrane of mitochondria such as Bnip3, Nix/Bnip3L, Bcl2L13, and FUNDC1, in mammalian cells. These receptors directly interact with Atg8 family proteins on the surface of mitochondria via their LIR motif and recruit autophagic machinery to damaged mitochondria [26–30]. It has been reported that NIX/Bnip3L is highly induced at the late stage of erythrocyte maturation and mediates autophagic degradation of mitochondria [28, 31, 32]. In addition, NIX/Bnip3L, BNIP3, and FUNDC1 are involved in hypoxia-induced mitophagy [33].

Mitophagy plays a critical role in controlling mitochondrial quality and quantity in response to the energy demand and other cellular or mitochondrial physiological condition. Besides described so far, mitophagy is induced by iron chelation in a BNIP3- and Parkin-independent manner [34], and contributes to paternal mitochondria elimination in *Caenorhabditis elegans* embryos [35, 36]. Recent progress in mitophagy studies has revealed a part of the molecular mechanism of mitophagy. However, there are many questions still to be addressed. In this book, we describe about currently established experimental protocols in mitophagy studies. In Part I, we focus to describe on PINK1/Parkin-mediated mitophagy specific methodology. In reference [37], Shigeto Sato and his colleagues describe about various protocols for induction of PINK1/Parkin-mediated mitophagy in cultured cells. In the subsequent chapters, we argue the detection methods for PINK1/Parkin-mediated mitophagy by immunofluorescent microscopy (by Shinji Saiki and his colleagues in reference [38]) and by correlative electron microscopy (by Chieko Kishi-Itakura and her colleagues in reference [39]). In addition, Yuzuru Imai and his colleagues mention PINK1/Parkin deficient Drosophila as an in vivo model of PINK1/Parkin-mediated mitophagy deficiency in reference [40]. Wado Akamatsu and his colleagues provide detailed protocols for the assessment of PINK1/Parkin-mediated mitophagy in iPS cells-derived neuron in reference [41]. In Part II, we focus on experimental protocols for the assessment of mitophagy in yeast using fluorescence microscopy and western blotting (Koji Okamoto and his colleague in reference [42]) and using MitoPho8Δ60 assay (by Daniel J. Klionsky and his colleagues in reference [43]). In addition, Tomotake Kanki and his colleague mention about genome wide screening for mitophagy-deficient mutants in reference [44]. In Part III, we argue about methodology for Nix/Bnip3L-dependent mitophagy. Ivana Novak and her colleagues argue about the assessment of Bnip3 and Bnip3L/Nix-dependent mitophagy by flow cytometric analysis in reference [45]. Du Feng and his colleagues describe about miRNA regulation of Nix in reference [46]. In the subsequent chapters, we describe on the methodologies about other types of mitophagy. In reference [47], Shigeomi Shimizu and his colleagues mention about monitoring methods for Atg5-independent mitophagy. In reference [48], Miyuki Sato and her colleague provide protocols to monitor allophagy, a type of mitophagy that degrades paternally inherited mitochondria and their mitochondrial DNA in *C. elegans* embryos. In reference [49], Tomotake Kanki and his colleague argue about hypoxia-induced and iron-depletion-induced mitophagy in mammalian cells. In reference [50], Nektarios Tavernarakis and his colleagues mention about methodology for mitophagy in *C. elegans* during aging.

References

1. Bratic A, Larsson NG (2013) The role of mitochondria in aging. J Clin Invest 123 (3):951–957. doi:10.1172/JCI64125
2. Lezi E, Swerdlow RH (2012) Mitochondria in neurodegeneration. Adv Exp Med Biol 942:269–286. doi:10.1007/978-94-007-2869-1_12
3. Pagano G, Talamanca AA, Castello G, Cordero MD, d'Ischia M, Gadaleta MN, Pallardo FV, Petrovic S, Tiano L, Zatterale A (2014) Oxidative stress and mitochondrial dysfunction across broad-ranging pathologies: toward mitochondria-targeted clinical strategies. Oxidative Med Cell Longev 2014:541230. doi:10.1155/2014/541230
4. Szendroedi J, Frossard M, Klein N, Bieglmayer C, Wagner O, Pacini G, Decker J, Nowotny P, Muller M, Roden M (2012) Lipid-induced insulin resistance is not mediated by impaired transcapillary transport of insulin and glucose in humans. Diabetes 61(12):3176–3180. doi:10.2337/db12-0108
5. Wallace DC (2012) Mitochondria and cancer. Nat Rev Cancer 12(10):685–698. doi:10.1038/nrc3365
6. Mizushima N, Komatsu M (2011) Autophagy: renovation of cells and tissues. Cell 147 (4):728–741. doi:10.1016/j.cell.2011.10.026
7. Galluzzi L, Pietrocola F, Levine B, Kroemer G (2014) Metabolic control of autophagy. Cell 159(6):1263–1276. doi:10.1016/j.cell.2014.11.006
8. Green DR, Levine B (2014) To be or not to be? How selective autophagy and cell death govern cell fate. Cell 157(1):65–75. doi:10.1016/j.cell.2014.02.049
9. Randow F, Youle RJ (2014) Self and nonself: how autophagy targets mitochondria and bacteria. Cell Host Microbe 15(4):403–411. doi:10.1016/j.chom.2014.03.012
10. Rogov V, Dotsch V, Johansen T, Kirkin V (2014) Interactions between autophagy receptors and ubiquitin-like proteins form the molecular basis for selective autophagy. Mol Cell 53(2):167–178. doi:10.1016/j.molcel.2013.12.014
11. Stolz A, Ernst A, Dikic I (2014) Cargo recognition and trafficking in selective autophagy. Nat Cell Biol 16(6):495–501. doi:10.1038/ncb2979
12. Narendra D, Walker JE, Youle R (2012) Mitochondrial quality control mediated by PINK1 and Parkin: links to parkinsonism. Cold Spring Harb Perspect Biol 4(11). doi:10.1101/cshperspect.a011338
13. Jin SM, Lazarou M, Wang C, Kane LA, Narendra DP, Youle RJ (2010) Mitochondrial membrane potential regulates PINK1 import and proteolytic destabilization by PARL. J Cell Biol 191(5):933–942. doi:10.1083/jcb.201008084
14. Lazarou M, Jin SM, Kane LA, Youle RJ (2012) Role of PINK1 binding to the TOM complex and alternate intracellular membranes in

recruitment and activation of the E3 ligase Parkin. Dev Cell 22(2):320–333. doi:10.1016/j.devcel.2011.12.014

15. Kane LA, Lazarou M, Fogel AI, Li Y, Yamano K, Sarraf SA, Banerjee S, Youle RJ (2014) PINK1 phosphorylates ubiquitin to activate Parkin E3 ubiquitin ligase activity. J Cell Biol 205(2):143–153. doi:10.1083/jcb.201402104
16. Kazlauskaite A, Kondapalli C, Gourlay R, Campbell DG, Ritorto MS, Hofmann K, Alessi DR, Knebel A, Trost M, Muqit MM (2014) Parkin is activated by PINK1-dependent phosphorylation of ubiquitin at Ser65. Biochem J 460(1):127–139. doi:10.1042/BJ20140334
17. Koyano F, Okatsu K, Kosako H, Tamura Y, Go E, Kimura M, Kimura Y, Tsuchiya H, Yoshihara H, Hirokawa T, Endo T, Fon EA, Trempe JF, Saeki Y, Tanaka K, Matsuda N (2014) Ubiquitin is phosphorylated by PINK1 to activate parkin. Nature 510(7503):162–166. doi:10.1038/nature13392
18. Ordureau A, Sarraf SA, Duda DM, Heo JM, Jedrychowski MP, Sviderskiy VO, Olszewski JL, Koerber JT, Xie T, Beausoleil SA, Wells JA, Gygi SP, Schulman BA, Harper JW (2014) Quantitative proteomics reveal a feedforward mechanism for mitochondrial PARKIN translocation and ubiquitin chain synthesis. Mol Cell 56(3):360–375. doi:10.1016/j.molcel.2014.09.007
19. Geisler S, Holmstrom KM, Skujat D, Fiesel FC, Rothfuss OC, Kahle PJ, Springer W (2010) PINK1/Parkin-mediated mitophagy is dependent on VDAC1 and p62/SQSTM1. Nat Cell Biol 12(2):119–131. doi:10.1038/ncb2012
20. Lazarou M, Narendra DP, Jin SM, Tekle E, Banerjee S, Youle RJ (2013) PINK1 drives Parkin self-association and HECT-like E3 activity upstream of mitochondrial binding. J Cell Biol 200(2):163–172. doi:10.1083/jcb.201210111
21. Yoshii SR, Kishi C, Ishihara N, Mizushima N (2011) Parkin mediates proteasome-dependent protein degradation and rupture of the outer mitochondrial membrane. J Biol Chem 286(22):19630–19640. doi:10.1074/jbc.M110.209338
22. Lazarou M, Sliter DA, Kane LA, Sarraf SA, Wang C, Burman JL, Sideris DP, Fogel AI, Youle RJ (2015) The ubiquitin kinase PINK1 recruits autophagy receptors to induce mitophagy. Nature 524(7565):309–314. doi:10.1038/nature14893
23. Heo JM, Ordureau A, Paulo JA, Rinehart J, Harper JW (2015) The PINK1-PARKIN mitochondrial Ubiquitylation pathway drives a program of OPTN/NDP52 recruitment and TBK1 activation to promote mitophagy. Mol Cell 60(1):7–20. doi:10.1016/j.molcel.2015.08.016
24. Kanki T, Wang K, Cao Y, Baba M, Klionsky DJ (2009) Atg32 is a mitochondrial protein that confers selectivity during mitophagy. Dev Cell 17(1):98–109. doi:10.1016/j.devcel.2009.06.014
25. Okamoto K, Kondo-Okamoto N, Ohsumi Y (2009) Mitochondria-anchored receptor Atg32 mediates degradation of mitochondria via selective autophagy. Dev Cell 17(1):87–97. doi:10.1016/j.devcel.2009.06.013
26. Hanna RA, Quinsay MN, Orogo AM, Giang K, Rikka S, Gustafsson AB (2012) Microtubule-associated protein 1 light chain 3 (LC3) interacts with Bnip3 protein to selectively remove endoplasmic reticulum and mitochondria via autophagy. J Biol Chem 287(23):19094–19104. doi:10.1074/jbc.M111.322933
27. Zhu Y, Massen S, Terenzio M, Lang V, Chen-Lindner S, Eils R, Novak I, Dikic I, Hamacher-Brady A, Brady NR (2013) Modulation of serines 17 and 24 in the LC3-interacting region of Bnip3 determines pro-survival mitophagy versus apoptosis. J Biol Chem 288(2):1099–1113. doi:10.1074/jbc.M112.399345
28. Novak I, Kirkin V, McEwan DG, Zhang J, Wild P, Rozenknop A, Rogov V, Lohr F, Popovic D, Occhipinti A, Reichert AS, Terzic J, Dotsch V, Ney PA, Dikic I (2010) Nix is a selective autophagy receptor for mitochondrial clearance. EMBO Rep 11(1):45–51. doi:10.1038/embor.2009.256
29. Murakawa T, Yamaguchi O, Hashimoto A, Hikoso S, Takeda T, Oka T, Yasui H, Ueda H, Akazawa Y, Nakayama H, Taneike M, Misaka T, Omiya S, Shah AM, Yamamoto A, Nishida K, Ohsumi Y, Okamoto K, Sakata Y, Otsu K (2015) Bcl-2-like protein 13 is a mammalian Atg32 homologue that mediates mitophagy and mitochondrial fragmentation. Nat Commun 6:7527. doi:10.1038/ncomms8527
30. Liu L, Feng D, Chen G, Chen M, Zheng Q, Song P, Ma Q, Zhu C, Wang R, Qi W, Huang L, Xue P, Li B, Wang X, Jin H, Wang J, Yang F, Liu P, Zhu Y, Sui S, Chen Q (2012) Mitochondrial outer-membrane protein FUNDC1 mediates hypoxia-induced mitophagy in mammalian cells. Nat Cell Biol 14(2):177–185. doi:10.1038/ncb2422
31. Schweers RL, Zhang J, Randall MS, Loyd MR, Li W, Dorsey FC, Kundu M, Opferman JT, Cleveland JL, Miller JL, Ney PA (2007) NIX is required for programmed mitochondrial

clearance during reticulocyte maturation. Proc Natl Acad Sci U S A 104(49):19500–19505. doi:10.1073/pnas.0708818104
32. Sandoval H, Thiagarajan P, Dasgupta SK, Schumacher A, Prchal JT, Chen M, Wang J (2008) Essential role for Nix in autophagic maturation of erythroid cells. Nature 454 (7201):232–235. doi:10.1038/nature07006
33. Li W, Zhang X, Zhuang H, Chen HG, Chen Y, Tian W, Wu W, Li Y, Wang S, Zhang L, Chen Y, Li L, Zhao B, Sui S, Hu Z, Feng D (2014) MicroRNA-137 is a novel hypoxia-responsive microRNA that inhibits mitophagy via regulation of two mitophagy receptors FUNDC1 and NIX. J Biol Chem 289(15):10691–10701. doi:10.1074/jbc.M113.537050
34. Schiavi A, Maglioni S, Palikaras K, Shaik A, Strappazzon F, Brinkmann V, Torgovnick A, Castelein N, De Henau S, Braeckman BP, Cecconi F, Tavernarakis N, Ventura N (2015) Iron-starvation-induced mitophagy mediates lifespan extension upon mitochondrial stress in *C. elegans*. Curr Biol 25(14):1810–1822. doi:10.1016/j.cub.2015.05.059
35. Al Rawi S, Louvet-Vallee S, Djeddi A, Sachse M, Culetto E, Hajjar C, Boyd L, Legouis R, Galy V (2011) Postfertilization autophagy of sperm organelles prevents paternal mitochondrial DNA transmission. Science 334 (6059):1144–1147. doi:10.1126/science.1211878
36. Sato M, Sato K (2011) Degradation of paternal mitochondria by fertilization-triggered autophagy in *C. elegans* embryos. Science 334 (6059):1141–1144. doi:10.1126/science.1210333
37. Sato S, Furuya N (2017) Induction of PINK1/Parkin-mediated mitophagy. Methods Mol Biol. doi:10.1007/7651_2017_7
38. Fujimaki M, Saiki S, Sasazawa Y, Ishikawa KI, Imamichi Y, Sumiyoshi K, Hattori N (2017) Immunocytochemical monitoring of PINK1/Parkin-mediated mitophagy in cultured cells. Methods Mol Biol. doi:10.1007/7651_2017_20
39. Kishi-Itakura C, Buss F (2017) The use of correlative light-electron microscopy (CLEM) to study PINK1/Parkin-mediated mitophagy. Methods Mol Biol. doi:10.1007/7651_2017_8
40. Inoshita T, Shiba-Fukushima K, Meng H, Hattori N, Imai Y (2017) Monitoring mitochondrial changes by alteration of the PINK1-Parkin signaling in Drosophila. Methods Mol Biol. doi:10.1007/7651_2017_9
41. Ishikawa KI, Yamaguchi A, Okano H, Akamatsu W (2017) Assessment of mitophagy in iPS cell-derived neurons. Methods Mol Biol. doi:10.1007/7651_2017_10
42. Nagumo S, Okamoto K (2017) Investigation of yeast mitophagy with fluorescence microscopy and western blotting. Methods Mol Biol. doi:10.1007/7651_2017_11
43. Yao Z, Liu X, Klionsky DJ (2017) MitoPho8Δ60 assay as a tool to quantitatively measure mitophagy activity. Methods Mol Biol. doi:10.1007/7651_2017_12
44. Furukawa K, Kanki T (2017) Mitophagy in yeast: a screen of mitophagy-deficient mutants. Methods Mol Biol. doi:10.1007/7651_2017_13
45. Šprung M, Dikic I, Novak I (2017) Flow cytometer monitoring of Bnip3- and Bnip3L/Nix-dependent mitophagy. Methods Mol Biol. doi:10.1007/7651_2017_14
46. Li W, Chen H, Li S, Lin G, Feng D (2017) Exploring MicroRNAs on NIX-dependent mitophagy. Methods Mol Biol. doi:10.1007/7651_2017_15
47. Arakawa S, Honda S, Torii S, Tsujioka M, Shimizu S (2017) Monitoring of Atg5-independent mitophagy. Methods Mol Biol. doi:10.1007/7651_2017_16
48. Sato M, Sato K (2017) Monitoring of paternal mitochondrial degradation in Caenorhabditis elegans. Methods Mol Biol. doi:10.1007/7651_2017_17
49. Yamashita SI, Kanki T (2017) Detection of hypoxia-induced and iron depletion-induced mitophagy in mammalian cells. Methods Mol Biol. doi:10.1007/7651_2017_19
50. Charmpilas N, Kounakis K, Tavernarakis N (2017) Monitoring mitophagy during aging in Caenorhabditis elegans. Methods Mol Biol. doi:10.1007/7651_2017_18

Methods in Molecular Biology (2018) 1759: 9–17
DOI 10.1007/7651_2017_7

Published online: 31 March 2017

Induction of PINK1/Parkin-Mediated Mitophagy

Shigeto Sato and Norihiko Furuya

Abstract

PINK1/Parkin mitophagy is a key mechanism to contribute mitochondrial quality control, and the defects are thought to be a cause of those Parkinson's disease onsets. Upon loss of mitochondrial membrane potential, PINK1 and Parkin are activated to promote the proteasomal degradation of mitochondrial outer membrane proteins and selective elimination of damaged mitochondria by autophagy. In this chapter, we describe the methods for induction of PINK1/Parkin-mediated mitophagy in tissue culture cell lines.

Keywords: Mitochondrial membrane potential, Parkin, PINK1

1 Introduction

PINK1/Parkin-mediated mitophagy is the best-characterized mechanism for elimination of damaged mitochondria by autophagy in mammalian cells [1]. Mutations in the mitochondrial Ser/Thr kinase PTEN-induced kinase 1 (PINK1) and the RING-HECT hybrid ubiquitin E3 ligase Parkin cause autosomal recessive juvenile Parkinson's disease [2–4]. The defects of PINK1/Parkin-mediated mitophagy are thought to be a cause of those Parkinson's disease onsets. In mitochondria, which lose the membrane potential or accumulate unfolded proteins, PINK1 is stabilized on the outer mitochondrial membrane (OMM) associating with TOMM complex [5, 6]. PINK1 recruits Parkin and activates latent E3 ligase activity of Parkin [7, 8]. Parkin activation leads to proteasomal degradation of OMM proteins and to selective autophagy of damaged mitochondria [9–11]. To induce PINK1/Parkin-mediated mitophagy experimentally, mitochondrial uncoupler, such as carbonyl cyanide m-chlorophenylhydrazone (CCCP), carbonyl cyanide-p-trifluoromethoxyphenylhydrazone (FCCP), or valinomycin, is used to depolarize mitochondrial membrane potential [1, 12]. Recently, cocktail of oligomycin, an inhibitor for oxidative phosphorylation (OXPHOS) complex V, and antimycin, an inhibitor for OXPHOS complex III, is also used for induction of PINK1/Parkin-mediated mitophagy [13, 14]. In addition, Yang W. Y. and his colleague reported that KillerRed-dMito, a mitochondrion-matrix-targeted photosensitizer, could induce PINK1/Parkin-mediated mitophagy [15].

In this chapter, we describe the experimental protocols for induction of PINK1/Parkin-mediated mitophagy using mitochondrial uncoupler, OXPHOS complex III and V inhibitor cocktail, and KillerRed-dMito. In addition to those protocols, we argue the methodology to confirm the loss of mitochondrial membrane potential and the activation of PINK1and Parkin.

2 Materials

2.1 Induction of PINK1/Parkin-Mediated Mitophagy by Mitochondrial Uncoupler

1. Appropriate cell line [e.g., HeLa cells, mouse embryonic fibroblast (MEF), HEK293 cells, SH-SY5Y cells].
2. Culture medium [e.g., for HeLa and HEK293 cells, Dulbecco's modified Eagle's medium (DMEM), 10 % fetal calf serum (FCS), and 100 U/ml penicillin/streptomycin; for MEF, DMEM, 10 % FCS, 100 U/ml penicillin/streptomycin, 1× nonessential amino acid (Gibco), 1 mM sodium pyruvate, and 0.0007 % 2-mercaptoethanol; for SH-SY5Y cells, DMEM, 10 % FCS, 100 U/ml penicillin/streptomycin, 2 mM L-glutamine, 1× nonessential amino acid (Gibco), and 1 mM sodium pyruvate].
3. Mammalian expression plasmid of Parkin [e.g., pEGFP-Parkin WT (Addgene #45875), pMXs-IP HA-Parkin (Addgene #38248), mCherry-Parkin (Addgene #23956), or equivalent].
4. Transfection reagent (e.g., Lipofectamine2000) or appropriate transfection device.
5. CCCP.
6. Valinomycin.
7. Oligomycin.
8. Antimycin A.

2.2 Confirmation of Mitochondrial Depolarization Using Fluorescent Dye

1. JC-1 (Molecular Probes #T3168).
2. Tetramethylrhodamine, ethyl ester (TMRE, Molecular Probes #T669).
3. MitoTracker Red CMXRos (Molecular Probes #M7512).

2.3 Confirmation of PINK1 Stabilization by Western Blotting

1. Lysis buffer [e.g., RIPA buffer (50 mM Tris–HCl, pH 7.4, 150 mM NaCl, 1 mM EDTA, 1 % Triton X-100, 1 % sodium deoxycholate, 0.1 % SDS) or equivalent].
2. Anti-PINK1 antibody (Novus #BC100-494 or Cell Signaling Technologies #6946).

2.4 Confirmation of Parkin Translocation to Mitochondria by Immunofluorescence Microscopy

1. Mammalian expression plasmid of Parkin (fluorescent protein-tag version is recommended), as described in Section 2.1.
2. Anti-epitope tag antibody (if necessary, e.g., Anti-HA).
3. Antibody against mitochondrial marker [e.g., anti-Tom20 (Santa Cruz Biotechnology #sc-11415, BD Bioscience #612278), anti-Tim23 (BD Bioscience #611222), anti-UQCRC1/Complex III Core I (Thermo Fisher Scientific #459140; Proteintech #21705-1-AP), or equivalent].
4. Fluorescence-conjugate secondary antibody (e.g., Alexa Fluor 488, 594, or 647 labeled anti-rabbit or anti-mouse antibody).
5. 16 % paraformaldehyde.
6. Triton X-100.
7. Blocking solution: PBS containing 4 % bovine serum albumin and 1 % goat serum.
8. Anti-fade mounting medium [e.g., VECTASHIELD Mounting Medium (VECTASHIELD #H-1000), ProLong Diamond anti-fade reagent (Thermo Fisher Scientific #P36961), or equivalent].
9. Colorless nail polish.

2.5 Confirmation of Parkin Translocation to Mitochondria by Cell Fractionation

1. Dithiobis [succinimidyl propionate] (DSP, Thermo Fisher Scientific #22585 or equivalent).
2. Glycine.
3. Chappell-Perry buffer: 150 mM KCl, 20 mM HEPES-NaOH pH 8.1, 5 mM $MgCl_2$, protease and phosphatase inhibitor cocktail (Roche #11873580001, #04906837001, respectively).
4. Anti-Parkin antibody (PRK8, Cell Signaling Technologies #4211).

2.6 Induction of PINK1/Parkin-Mediated Mitophagy Using KillerRed-dMito

1. HeLa cells.
2. Culture medium for HeLa cells: DMEM, 10 % FCS, and 100 U/ml penicillin/streptomycin.
3. KillerRed-dMito (Evrogen #FP964).
4. Fluorescent protein-tagged mammalian expression plasmid of Parkin [e.g., pEGFP-Parkin WT (Addgene #45875) or YFP-Parkin expression plasmid (Addgene #23955)].
5. Phenol red-free medium (e.g., Thermo Fisher Scientific #21063029 or equivalent).

3 Methods

3.1 Induction of PINK1/Parkin-Mediated Mitophagy by Mitochondrial Uncoupler

1. Transfect appropriate mammalian expression plasmid of Parkin into cells (*see* **Note 1**) using transfection reagent.
2. 24–48 h after transfection, incubated cells with or without CCCP (10 μM for HeLa cells, 30 μM for MEF), valinomycin (10 μM for HeLa cells), or cocktail of oligomycin and antimycin A (OA; 10 μM and 5 μM, respectively, for HeLa cells) for 1–24 h (*see* **Note 2**).

3.2 Confirmation of Mitochondrial Depolarization Using Fluorescent Dye

1. Incubate Parkin-expressing cells cultured in cellware for imaging (e.g., glass bottom dishes, chamber slides, plates for imaging, etc.), with or without CCCP, valinomycin, or OA to depolarize mitochondria, as described above.
2. Add 10 μg/ml JC1, 500 nM TMRE, or 50 nM MitoTracker CMXRos to cells at the final 20 min of uncoupler treatment (*see* **Notes 3** and **4**).
3. Observe the cells under confocal laser microscope with appropriate excitation and emission filters (JC1 (Ex. 535/Em. 590 represents healthy mitochondria, Ex. 485/Em. 505 represents depolarized mitochondria), TMRE (Ex. 543/Em. 573), MitoTracker CMXRos (Ex. 579/Em. 599), Fig. 1).

3.3 Confirmation of PINK1 Stabilization by Western Blotting

1. Incubate cells in 6-well plate or equivalent cellware with or without mitochondrial uncoupler as described in Section 3.1.
2. Wash cells with PBS once, then lyse cells in Lysis buffer for 30 min at 4 °C.
3. Harvest cell lysate using cell scraper.
4. To remove cell debris, centrifuge the lysate at 12,000 × *g* for 5 min at 4 °C.
5. Measure the protein concentration of lysate and prepare sample for SDS-PAGE with sample buffer.
6. Samples are subjected to Western blotting using anti-PINK1 antibody (Fig. 2; *see* **Note 5**).

3.4 Confirmation of Parkin Translocation to Mitochondria by Immunofluorescence Microscopy

1. Transfect mammalian expression plasmid of Parkin into cells in imaging cellware.
2. 24–48 h after transfection, incubate cells with or without mitochondrial uncoupler, as described in Section 3.1.
3. Remove media and fix the cells with PBS containing 4 % paraformaldehyde for 15 min at room temperature (RT).
4. Permeabilize the cells with PBS containing 0.2 % Triton X-100 for 5 min at RT.
5. Wash the cells three times for 10 min with PBS.

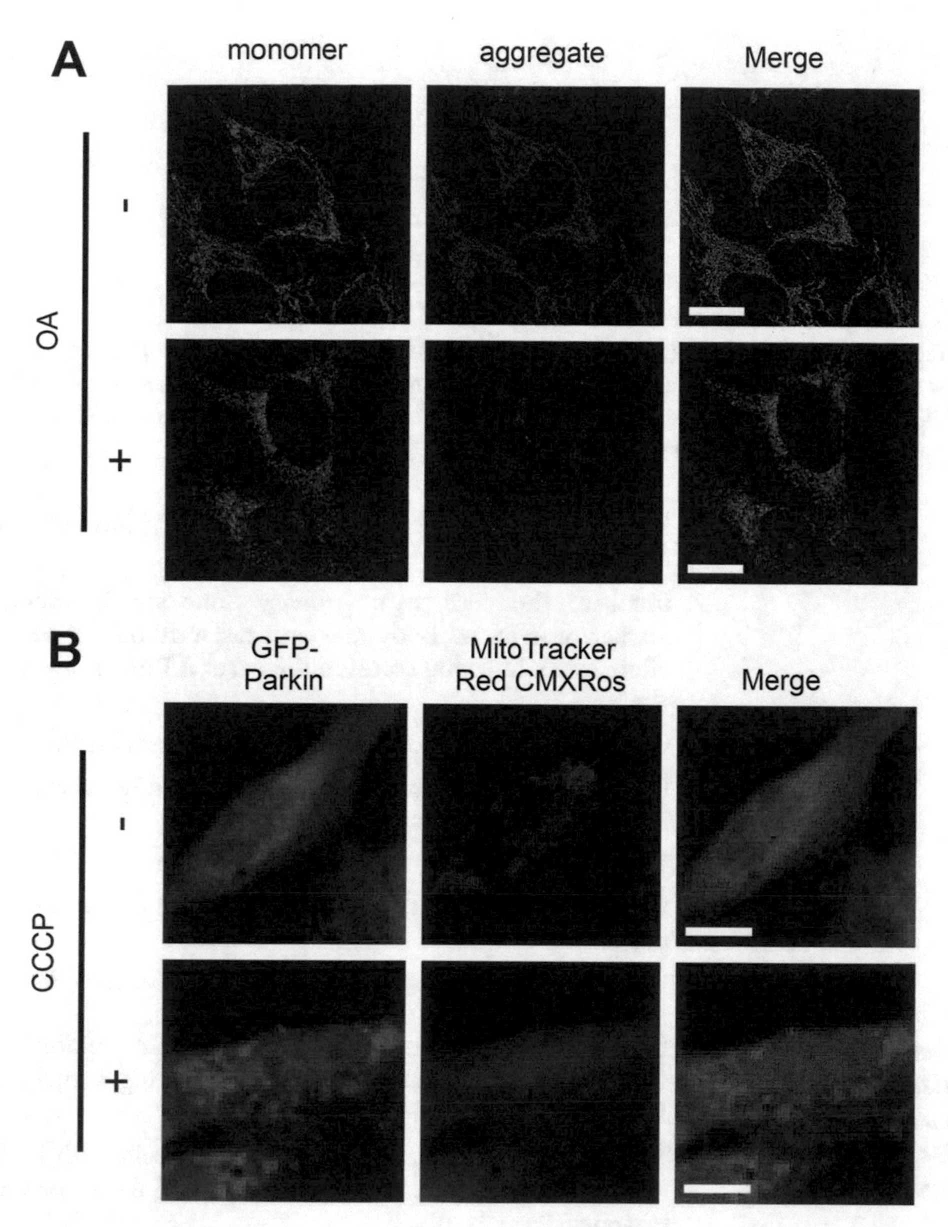

Fig. 1 Assessment of mitochondrial membrane potential using fluorescent dyes. (**a**) HeLa cells cultured with or without oligomycin/antimycin A (OA) for 4 h were treated with JC-1. Green-fluorescent monomer and red-fluorescent J-aggregate show depolarized membrane potentials and hyperpolarized membrane potentials, respectively. Scale bar, 10 μm. (**b**) HeLa cells expressing GFP-Parkin were treated with MitoTracker CMXRos in the presence or absence of CCCP treatment. While MitoTracker dye showed "mitochondrial pattern" in the absence of CCCP, MitoTracker staining was spread to whole cytoplasmic area in the presence of CCCP. Scale bar, 10 μm

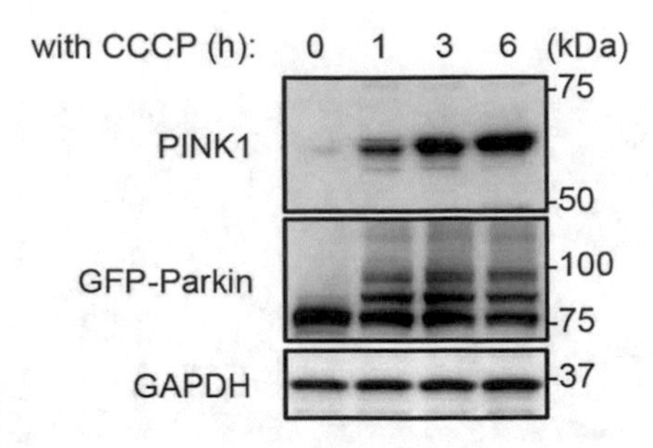

Fig. 2 Representative Western blotting using anti-PINK1 antibody in CCCP treated cells. HeLa cells expressing GFP-Parkin cells were incubated with CCCP for 0, 1, 2, and 3 h. Cell lysate was subjected to SDS-PAGE followed by Western blotting using anti-PINK1 (Novus #BC100-494), anti-Parkin (Cell Signaling Technologies #4211), and anti-GAPDH as an internal control

6. For blocking, incubate the cells with blocking solution for 30–60 min at RT.
7. Incubate the cells with primary antibody (mitochondrial marker protein antibody and anti-tag antibody, if necessary) diluted with blocking solution for 1 h at RT or for overnight at 4 °C.
8. Wash the cells three to five times for 10 min with PBS.
9. Incubate the cells with secondary antibody diluted with blocking solution for 1 h at RT.
10. Wash the cells three to five times for 10 min with PBS.
11. Overlay coverslips with mounting medium and seal with nail polish.
12. Observe the cells using confocal laser microscope (Fig. 3).

3.5 Confirmation of Parkin Translocation to Mitochondria by Cell Fractionation

1. Incubate Parkin-expressing cells with CCCP, valinomycin, or OA; subsequently treat with in PBS containing 1 mM DSP for 1 h on ice (*see* **Note 6**).
2. To terminate cross-linking reaction, wash cells with PBS containing 10 mM glycine for three times, and then resuspend in Chappell-Perry buffer.
3. Disrupt the cells by five passages through a 25-gauge needle with 1-ml syringe.
4. To remove cell debris, centrifuge the homogenate at 1,000 × *g* for 7 min at 4 °C.
5. To separate mitochondrion-rich fraction and cytosol, centrifuge the supernatant at 10,000 × *g* for 10 min at 4 °C.
6. Wash the precipitate (mitochondrion-rich fraction) with Chappell-Perry buffer for three times to remove contamination of cytosol, and resuspend with sample buffer of SDS-PAGE.

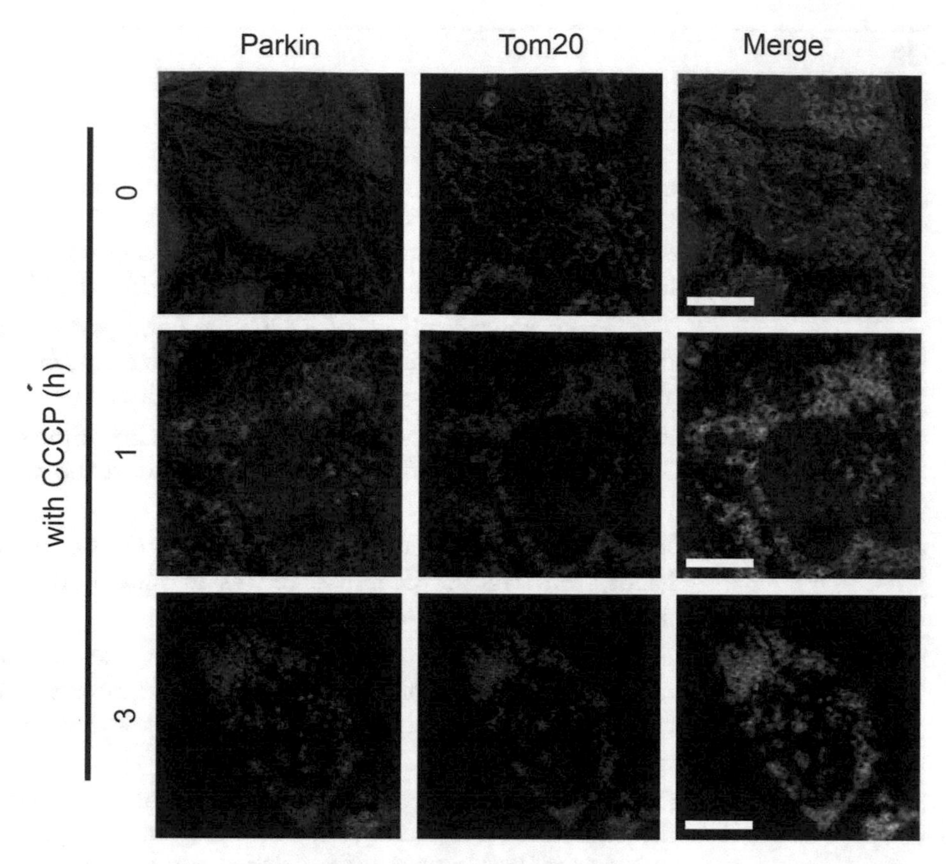

Fig. 3 Parkin translocation to mitochondria by mitochondrial uncoupler treatment. HeLa cells transfected with GFP-Parkin were incubated with CCCP for 0, 1, or 3 h. *Left* column, GFP-Parkin; *center* column, Alexa594 anti-Tom20, a mitochondrial marker, immunostaining; *right* column, overlay of GFP and Alexa594 staining. *White* color represents colocalization of Parkin and Tom20

Also resuspend the 10,000 × *g* supernatant (cytosol-rich fraction) with sample buffer of SDS-PAGE.

7. Samples are subjected to Western blotting using anti-Parkin antibody.

3.6 Induction of PINK1/Parkin-Mediated Mitophagy Using KillerRed-dMito

1. Co-express KillerRed-dMito and fluorescent protein-tagged (GFP or YFP) Parkin into HeLa cells.
2. 24–48 h after transfection, transfer cells into phenol-red-free medium.
3. Under confocal microscope, irradiate 50-μW 559-nm laser light through a ~10-μm × 10-μm region in cells with KillerRed-dMito and Parkin expression.
4. Capture images for 180 min with 1- to 5-min intervals (*see* **Note 7**).

4 Notes

1. HeLa cells and MEF do not have functional level of endogenous Parkin expression. Therefore, when HeLa cells or MEF are used for the study of PINK1/Parkin-mediated mitophagy, exogenous Parkin expression is required. Procedure of Section 3.1 is not required for studies using cells with endogenous Parkin expression (e.g., HEK293 or SHSY5Y cells).
2. The exact concentration of CCCP, valinomycin, and OA depends on the cell lines being used. Exact concentrations should be determined on an individual cell lines.
3. The exact concentration and effective incubation time of JC1, TMRE, and MitoTracker CMXRos depend on the cell lines being used.
4. JC1 and TMRE can be used only in living cell. On the other hands, MitoTracker CMXRos retains in mitochondria after paraformaldehyde fixation.
5. Endogenous PINK1 is rarely detected in the absence of mitochondrial uncoupler. More than 1-h treatment of mitochondrial uncoupler increases PINK1 expression detected as ~60-kDa protein.
6. Association of Parkin to depolarized mitochondria is very fragile. Therefore cross-linking reaction using DSP is required for retention of Parkin on mitochondrion-enriched fraction.
7. To maintain environment of HeLa cell culture during time-lapse imaging, confocal microscope with a microscope stage incubation chamber, which controls temperature, atmosphere, and humidity, is recommended for observation of cells.

References

1. Narendra D, Walker JE, Youle R (2012) Mitochondrial quality control mediated by PINK1 and Parkin: links to Parkinsonism. Cold Spring Harb Perspect Biol 4(11). doi:10.1101/cshperspect.a011338
2. Pickrell AM, Youle RJ (2015) The roles of PINK1, Parkin, and mitochondrial fidelity in Parkinson's disease. Neuron 85(2):257–273. doi:10.1016/j.neuron.2014.12.007
3. van der Merwe C, Jalali Sefid Dashti Z, Christoffels A, Loos B, Bardien S (2015) Evidence for a common biological pathway linking three Parkinson's disease-causing genes: Parkin, PINK1 and DJ-1. Eur J Neurosci 41(9):1113–1125. doi:10.1111/ejn.12872
4. Lazarou M, Narendra DP, Jin SM, Tekle E, Banerjee S, Youle RJ (2013) PINK1 drives Parkin self-association and HECT-like E3 activity upstream of mitochondrial binding. J Cell Biol 200(2):163–172. doi:10.1083/jcb.201210111
5. Jin SM, Lazarou M, Wang C, Kane LA, Narendra DP, Youle RJ (2010) Mitochondrial membrane potential regulates PINK1 import and proteolytic destabilization by PARL. J Cell Biol 191(5):933–942. doi:10.1083/jcb.201008084
6. Lazarou M, Jin SM, Kane LA, Youle RJ (2012) Role of PINK1 binding to the TOM complex and alternate intracellular membranes in recruitment and activation of the E3 ligase Parkin. Dev Cell 22(2):320–333. doi:10.1016/j.devcel.2011.12.014

7. Matsuda N, Sato S, Shiba K, Okatsu K, Saisho K, Gautier CA, Sou YS, Saiki S, Kawajiri S, Sato F, Kimura M, Komatsu M, Hattori N, Tanaka K (2010) PINK1 stabilized by mitochondrial depolarization recruits Parkin to damaged mitochondria and activates latent Parkin for mitophagy. J Cell Biol 189(2):211–221. doi:10.1083/jcb.200910140
8. Narendra DP, Jin SM, Tanaka A, Suen DF, Gautier CA, Shen J, Cookson MR, Youle RJ (2010) PINK1 is selectively stabilized on impaired mitochondria to activate Parkin. PLoS Biol 8(1):e1000298. doi:10.1371/journal.pbio.1000298
9. Yoshii SR, Kishi C, Ishihara N, Mizushima N (2011) Parkin mediates proteasome-dependent protein degradation and rupture of the outer mitochondrial membrane. J Biol Chem 286(22):19630–19640. doi:10.1074/jbc.M110.209338
10. Geisler S, Holmstrom KM, Skujat D, Fiesel FC, Rothfuss OC, Kahle PJ, Springer W (2010) PINK1/Parkin-mediated mitophagy is dependent on VDAC1 and p62/SQSTM1. Nat Cell Biol 12(2):119–131. doi:10.1038/ncb2012
11. Gegg ME, Cooper JM, Chau KY, Rojo M, Schapira AH, Taanman JW (2010) Mitofusin 1 and mitofusin 2 are ubiquitinated in a PINK1/Parkin-dependent manner upon induction of mitophagy. Hum Mol Genet 19(24):4861–4870. doi:10.1093/hmg/ddq419
12. Rakovic A, Grunewald A, Seibler P, Ramirez A, Kock N, Orolicki S, Lohmann K, Klein C (2010) Effect of endogenous mutant and wild-type PINK1 on Parkin in fibroblasts from Parkinson disease patients. Hum Mol Genet 19(16):3124–3137. doi:10.1093/hmg/ddq215
13. Heo JM, Ordureau A, Paulo JA, Rinehart J, Harper JW (2015) The PINK1-PARKIN mitochondrial Ubiquitylation pathway drives a program of OPTN/NDP52 recruitment and TBK1 activation to promote mitophagy. Mol Cell 60(1):7–20. doi:10.1016/j.molcel.2015.08.016
14. Lazarou M, Sliter DA, Kane LA, Sarraf SA, Wang C, Burman JL, Sideris DP, Fogel AI, Youle RJ (2015) The ubiquitin kinase PINK1 recruits autophagy receptors to induce mitophagy. Nature 524(7565):309–314. doi:10.1038/nature14893
15. Yang JY, Yang WY (2011) Spatiotemporally controlled initiation of Parkin-mediated mitophagy within single cells. Autophagy 7(10):1230–1238. doi:10.4161/auto.7.10.16626

Methods in Molecular Biology (2018) 1759: 19–27
DOI 10.1007/7651_2017_20

Published online: 31 March 2017

Immunocytochemical Monitoring of PINK1/Parkin-Mediated Mitophagy in Cultured Cells

Motoki Fujimaki, Shinji Saiki, Yukiko Sasazawa, Kei-Ichi Ishikawa, Yoko Imamichi, Katsuhiko Sumiyoshi, and Nobutaka Hattori

Abstract

Both PINK1 and parkin are the responsible genes (PARK6 and PARK2, respectively) for familial early-onset Parkinson's disease (PD). Several lines of evidences have suggested that mitochondrial dysfunction would be associated with PD pathogenesis. Lewy body, one of PD pathological hallmarks, contains alpha-synuclein, a familial PD (PARK1/4)-gene product, which is eliminated by macroautophagy, while PINK1 and parkin coordinately mediate mitophagy (hereafter called as PINK1/parkin-mediated mitophagy) reported firstly by Youle's group. The mitochondrial quality control system is specific for elimination of damaged mitochondria especially in the loss of mitochondrial membrane potential induced by treatment with mitochondrial uncoupler like CCCP or FCCP. In this chapter, we summarized immunocytochemical methods to monitor the PINK1/parkin-mediated mitophagy using cultured cells.

Keywords: Familial Parkinson's disease, PINK1, Parkin, Mitophagy, Immunocytochemistry, Parkinson's disease, Membrane potential, Mitochondrial uncoupler

1 Introduction

Two Parkinson's disease (PD) genes, PTEN-induced putative kinase 1 (PINK1) and parkin, which are responsible for autosomal recessive PD, are associated with mitochondrial function and coordinately control mitophagy reported by others and us [1–6]. Loss of mitochondrial membrane potentials induces accumulation of PINK1 followed by parkin recruitment to their outer membrane. Also, only PINK1 accumulation by its exogenous overexpression can induce PINK1/parkin mitophagy in cells with enough expression of parkin. Although PINK1 is rapidly degraded by proteolysis in healthy mitochondria, it accumulates in the damaged mitochondria with loss of their membrane potential by escaping from rapid proteolysis by PARL and recruits parkin to the damaged mitochondria followed by their ubiquitination by parkin [7]. Those ubiquitinated mitochondria are engulfed by isolation membrane that finally fuse with lysosomes. Defective autophagy results in the accumulation of mitochondria with an abnormal membrane potential, which are more likely to release proapoptotic molecules. Thus, PINK1/parkin-mediated mitophagy is generally considered as a cytoprotective response to eliminate the damaged mitochondria.

In this chapter, we would like to summarize an experimental monitoring protocol for PINK1/parkin-mediated mitophagy using HeLa cells.

2 Materials

2.1 Cell Cultures and Reagents

1. Cell lines: HeLa cells.
2. Culture medium: Dulbecco's Modified Eagle's Medium (DMEM), 4 mM L-glutamine, 10% fetal bovine serum (FBS), and 100 U/ml penicillin/streptomycin.
3. Culture condition: 37 °C, 5% CO_2.
4. Transfection reagent: Lipofectamine 2000 Reagent.
5. 10 μM CCCP (carbonylcyanide m-chlorophenylhydrazone).
6. 5 μM valinomycin.
7. 50 μg/ml digitonin.
8. 1× phosphate-buffered saline (PBS).

2.2 Confocal Microscopy Analysis of Mitophagy

1. Inverted Zeiss confocal microscope LSM510 or LSM780 using ZEN software.
2. Mitochondrial staining: first antibody, anti-Tom20 antibodies (rabbit or mouse), and second antibody, anti-rabbit or mouse Alexa 594, or 633. For further detailed technical descriptions about the antibodies, please refer to Refs. [1, 2].
3. Autophagosome/autolysosome staining: first antibody, anti-LC3B (rabbit), and second antibody, anti-rabbit Alexa 594, or 633. For further detailed technical descriptions about the antibodies, please refer to Refs. [1, 2].
4. Plasmid DNAs (EGFP-parkin, pEGFP-empty vector, PINK1-FLAG, and FLAG empty vector) were described in our previous reports [1, 2].

3 Methods

HeLa cells with lack of functional parkin expression [8] are used with exogenous parkin overexpression in induction assay of PINK1/parkin-mediated mitophagy. No antibodies against endogenous PINK1 or parkin have been established for immunocytochemistry at steady-state conditions; both of them have been exogenously expressed as tagged plasmid DNAs (e.g., GFP,

FLAG). Two independent protocols, with or without treatment using mitochondrial uncoupler (CCCP or valinomycin), are provided in this section.

In order to monitor PINK1/parkin-mediated mitophagy precisely using immunocytochemistry, consecutive analysis of PINK1/parkin-mediated mitophagy for 24 h after transfection with CCCP/valinomycin treatment should be performed. A 48-h treatment with CCCP/valinomycin is too toxic against HeLa cells to monitor it because of excess of cell death induction. On the other hand, complete elimination of fluorescent mitochondria detected by confocal microscopy is observed in HeLa cells 48 h after transfection of both PINK1-FLAG and pEGFP-parkin, leading easy detection of cells with mitophagy induction.

3.1 Mitophagy Induction by Mitochondrial Depolarization Using Mitochondrial Uncoupler, Valinomycin

1. 2.5×10^5 HeLa cells were spitted into each well of the 6-well plate with a cover slip.
2. Incubated in normal DMEM overnight.
3. Cells were transfected with pEGFP-parkin/pEGFP-empty vector (1.0 μg/well) according to a protocol of Lipofectamine 2000 (*see* **Note 1**).
4. 24 h after transfection, cells were treated with 10 μM valinomycin or DMSO for 4 h or 24 h (*see* **Note 2**).
5. Cells were fixed with 4% paraformaldehyde for 20 min and permeabilized with 50 μg/ml digitonin in 1× PBS for 10 min followed by blocking with 1% BSA/10% FBS for 1 h (*see* **Note 3**, permeabilization is insufficient with other detergents).
6. Cells were washed three times with 1× PBS for 10 min stained with first antibodies including Tom20 (mitochondria) (×200), LC3B (autophagosome) (×100), or LAMP2 (late endosome or lysosome) (×50) overnight.
7. Cells were washed three times with 1× PBS for 10 min and incubated with corresponding second antibodies, anti-rabbit IgG tagged with Alexa Fluor 633 or anti-mouse IgG tagged with Alexa Fluor 594 for 1 h.
8. Cover slips were mounted with VECTASHIELD containing DAPI and observed using a confocal microscopy. The cover slips were embedded with VECTASHIELD and stained with DAPI, and images were acquired on a Zeiss LSM510 META confocal microscope (63× 1.4 NA) or a Zeiss LSM780 confocal microscope (63× 1.4 NA) at room temperature using ZEN software. Adobe Photoshop 7.0 was used for subsequent image processing (Fig. 1) (*see* **Note 4**).

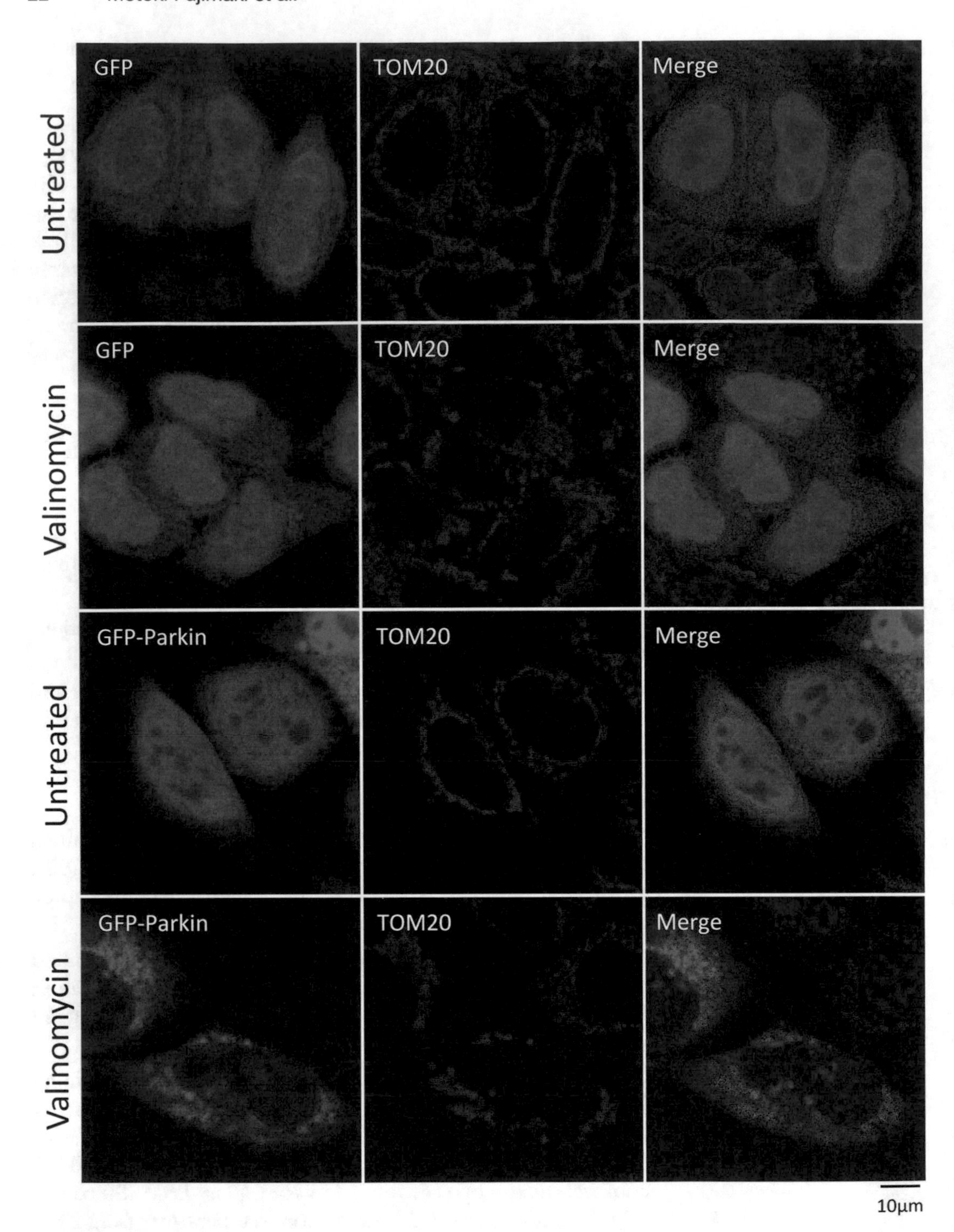

Fig. 1 HeLa cells expressing GFP or GFP-parkin (wild type) were treated with 5 μM valinomycin for 4 h followed by immunocytochemistry. In cells expressing GFP-parkin, valinomycin treatment changes the parkin localization to the aggregated mitochondria positive to Tom20 around the nucleus

3.2 Without Mitochondrial Uncoupler, Mitophagy Induction Initiation by Exogenously Overexpression of PINK1 as Well as Parkin

1. 2.5×10^5 HeLa cells were spitted into each well of the 6-well plate with a cover slip.
2. Incubated in normal DMEM overnight.
3. Cells were transfected with pEGFP-parkin/pEGFP-empty vector (0.3 μg/well) and PINK1-FLAG/FLAG empty vector (0.6 μg/well) according to a protocol of Lipofectamine 2000.
4. 24 or 48 h after transfection, cells were fixed with 4% paraformaldehyde and permeabilized with 50 μg/ml digitonin in 1× PBS for 10 min followed by blocking with 1% BSA/10% FBS for 1 h (*see* **Note 5**).
5. Cells were stained with first antibodies using anti-FLAG, Tom20 (mitochondria) (×200), LC3B (autophagosome) (×100), or LAMP2 (late endosome or lysosome) (×50) overnight.
6. Cells were washed three times with 1× PBS for 10 min and incubated with corresponding second antibodies, anti-rabbit IgG tagged with AlexaFluor 633 or anti-mouse IgG tagged with Alexa Fluor 594 for 1 h.
7. Cells were then mounted with VECTASHIELD containing DAPI and observed using a confocal microscopy. The cover slips were embedded with VECTASHIELD and stained with DAPI, and images were acquired on a Zeiss LSM510 META confocal microscope (63× 1.4 NA) or a Zeiss LSM780 confocal microscope (63× 1.4 NA) at room temperature using ZEN software. Adobe Photoshop 7.0 (Adobe Systems, Inc.) was used for subsequent image processing (Fig. 2a, b) (*see* **Note 6**).

4 Notes

1. Excess of transfected pEGFP-parkin plasmid DNAs easily aggregates mainly around the nucleus. To avoid this situation, the amount of pEGFP-parkin plasmid DNAs should be less than 0.5 μg/well (6-well plate).
2. Even 1 h after CCCP/valinomycin addition, PINK1/parkin-mediated mitophagy is triggered and observed as mitochondria morphological changes associated with PINK1 accumulation. Thus, if you would need to monitor its early steps, you should prepare samples with shorter treatment (e.g., 1–3 h) under the same condition.
3. Although we examined various conditions with other detergents to detect mitochondria, autophagosomes, autolysosomes, and/or lysosomes clearly, 10 μg/ml digitonin for

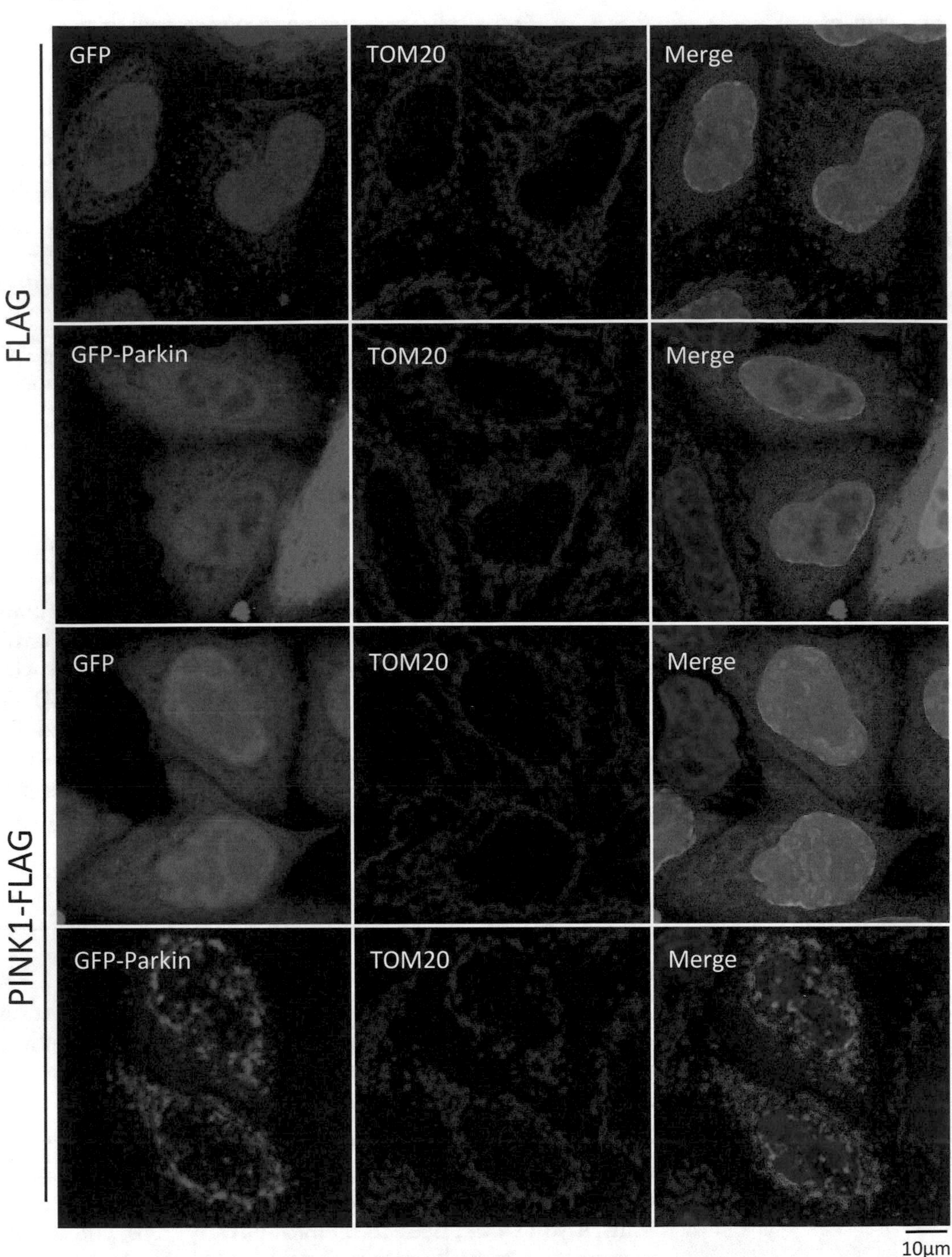

Fig. 2 Consecutive analysis of mitochondrial elimination by PINK1/parkin-mediated mitophagy by overexpression of wild-type PINK1 in combination with parkin. Immunocytochemistry of HeLa cells 24 or 48 h after transient co-overexpression of PINK1-WT and GFP-parkin. Aggregated mitochondria positive to GFP-parkin were detected only in cells overexpressing GFP-parkin as well as PINK1-FLAG 24 h after transfection (**a**). 48 h after transfection, some cells with markedly reduced mitochondria mildly positive to GFP-parkin immunofluorescence were occasionally observed (**b**)

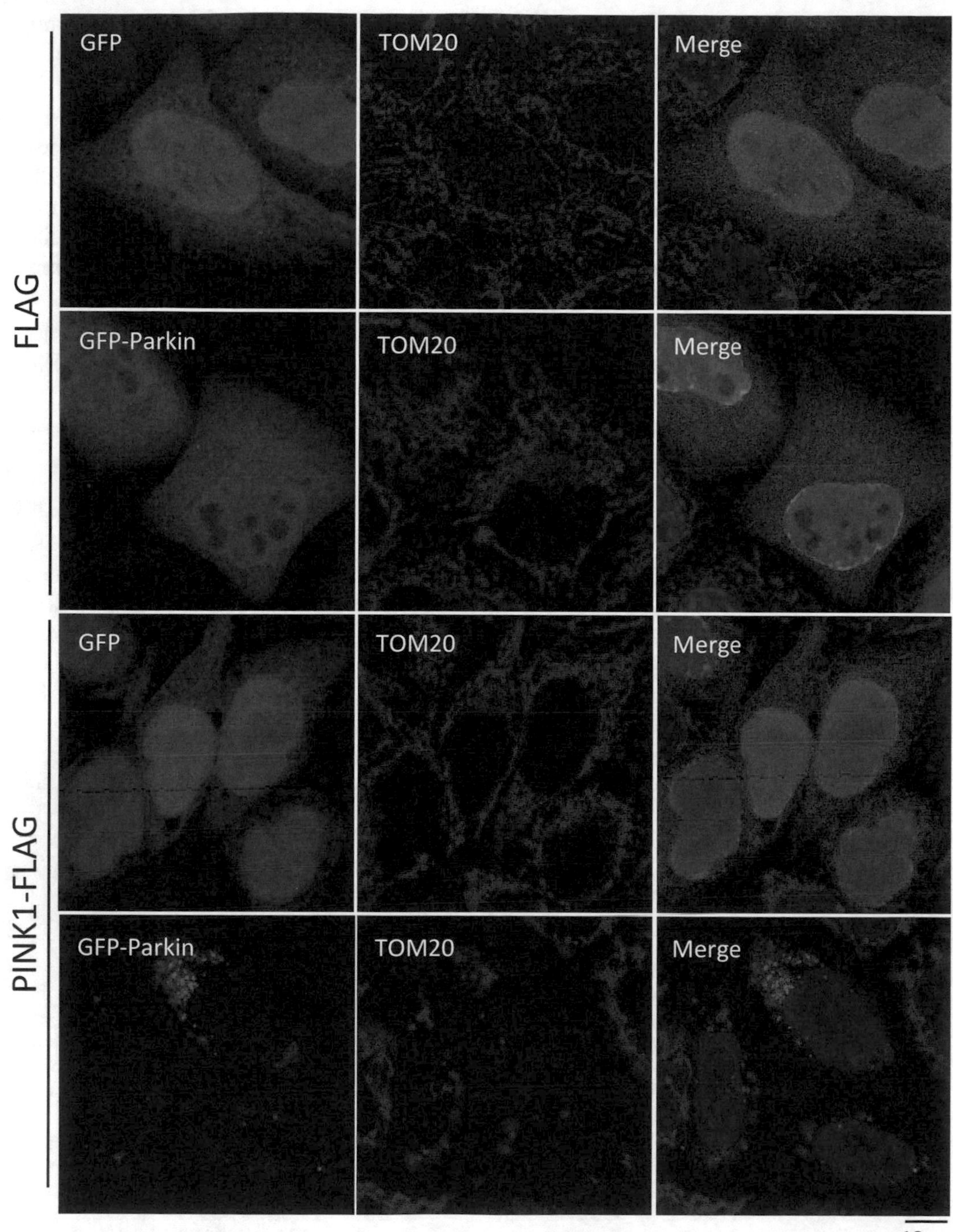

Fig. 2 (continued)

30 min incubation is the optima condition for HeLa cells compared to Triton X-100, Tween-20, and so on.

4. For quantitative analysis of mitophagosomes (aggregated mitochondria around the nucleus positive to LC3), you can add analysis with ImageJ "colocalization" plugin.
5. This protocol is described in detail in our report [1]. In this protocol, you do not have to lose mitochondrial membrane potentials for this mitophagy induction, suggesting that all we have to do for induction of PINK1/parkin-mediated mitophagy is to accumulate PINK1 excessing presenilin-associated rhomboid-like protein (PARL) protease capacity localizing in the mitochondrial intermembrane spaces.
6. For colocalization quantification, you can use ImageJ "colocalization" plugin according to our previous reports [9, 10].

Acknowledgment

We are grateful for the Grant-in-Aid for Principle Area (S.S., 25111007), the Grant-in-Aid for Scientific Research (B) (S.S., 15H04843), the Grant-in-Aid for Challenging Exploratory Research (S.S., 24659435), the Grant-in-Aid for JSPS Research Fellow (Y.S., 16 J40133), the Grant-in-Aid for Young Scientists (B) (K.I., JP16K19524), and the Grant (Y.I. JP16H00625) from Japan Society for the Promotion of Science and Subsidies to Private Schools.

Competing Financial Interests

The authors declare that they have no competing financial interests.

References

1. Kawajiri S, Saiki S, Sato S, Sato F, Hatano T, Eguchi H, Hattori N (2010) PINK1 is recruited to mitochondria with parkin and associates with LC3 in mitophagy. FEBS Lett 584(6):1073–1079. doi:10.1016/j.febslet.2010.02.016
2. Matsuda N, Sato S, Shiba K, Okatsu K, Saisho K, Gautier CA, Sou YS, Saiki S, Kawajiri S, Sato F, Kimura M, Komatsu M, Hattori N, Tanaka K (2010) PINK1 stabilized by mitochondrial depolarization recruits Parkin to damaged mitochondria and activates latent Parkin for mitophagy. J Cell Biol 189(2):211–221. doi:10.1083/jcb.200910140
3. Geisler S, Holmstrom KM, Skujat D, Fiesel FC, Rothfuss OC, Kahle PJ, Springer W (2010) PINK1/Parkin-mediated mitophagy is dependent on VDAC1 and p62/SQSTM1. Nat Cell Biol 12(2):119–131. doi:10.1038/ncb2012
4. Narendra DP, Jin SM, Tanaka A, Suen DF, Gautier CA, Shen J, Cookson MR, Youle RJ (2010) PINK1 is selectively stabilized on impaired mitochondria to activate Parkin. PLoS Biol 8(1):e1000298. doi:10.1371/journal.pbio.1000298
5. Vives-Bauza C, de Vries RL, Tocilescu M, Przedborski S (2010) PINK1/Parkin direct

mitochondria to autophagy. Autophagy 6 (2):315–316

6. Saiki S, Sato S, Hattori N (2012) Molecular pathogenesis of Parkinson's disease: update. J Neurol Neurosurg Psychiatry 83:430–436. doi:10.1136/jnnp-2011-301205
7. Youle RJ, Narendra DP (2011) Mechanisms of mitophagy. Nat Rev Mol Cell Biol 12(1):9–14. doi:10.1038/nrm3028
8. Denison SR, Wang F, Becker NA et al (2003) Alterations in the common fragile site gene Parkin in ovarian and other cancers. Oncogene 22:8370–8378
9. Saiki S, Sasazawa Y, Imamichi Y, Kawajiri S, Fujimaki T, Tanida I, Kobayashi H, Sato F, Sato S, Ishikawa K, Imoto M, Hattori N (2011) Caffeine induces apoptosis by enhancement of autophagy via PI3K/Akt/mTOR/p70S6K inhibition. Autophagy 7(2):176–187
10. Korolchuk VI, Saiki S, Lichtenberg M, Siddiqi FH, Roberts EA, Imarisio S, Jahreiss L, Sarkar S, Futter M, Menzies FM, O'Kane CJ, Deretic V, Rubinsztein DC (2011) Lysosomal positioning coordinates cellular nutrient responses. Nat Cell Biol 13(4):453–460. doi:10.1038/ncb2204

Methods in Molecular Biology (2018) 1759: 29–39
DOI 10.1007/7651_2017_8

Published online: 31 March 2017

The Use of Correlative Light-Electron Microscopy (CLEM) to Study PINK1/Parkin-Mediated Mitophagy

Chieko Kishi-Itakura and Folma Buss

Abstract

In this chapter we describe the use of correlative light-electron microscopy (CLEM) to study, in cultured cells, the turnover of damaged mitochondria by PINK1/Parkin-dependent mitophagy. CLEM combines the advantages of light microscopy, which allows to image and rapidly screen a large number of the cells, while electron microscopy provides high-resolution imaging of these selected cells and a detailed structural analysis of their cellular organelles. We describe in detail how to prepare the cell cultures for optimum preservation of their cellular ultrastructure for CLEM using the most suitable buffers, fixatives, and embedding resins. These protocols are applicable for detailed ultrastructural analysis in a wide variety of organisms and cells, ranging from prokaryotic bacteria to mammalian cells.

Keywords: Correlative light-electron microscopy (CLEM), Electron microscope, Mitophagy, Parkin

1 Introduction

Mitophagy is the selective process for degradation of damaged mitochondria by autophagy, which is essential for maintaining cellular homeostasis and survival. PTEN-induced putative kinase 1 (PINK1) and Parkin, a RING domain-containing E3 ubiquitin ligase, are mutated in many cases of autosomal recessive juvenile Parkinson's disease (PD) [1]. Recent studies in cultured cells have revealed that Parkin is important for mitochondrial quality control by initiating the degradation of damaged mitochondria. Narendra et al. [2] first demonstrated that Parkin translocates from the cytosol to depolarized mitochondria and triggers elimination of these damaged mitochondria by the autophagic process known as mitophagy. To initiate this process, tissue culture cell lines can be treated with the chemical uncoupler carbonyl cyanide *m*-chlorophenyl hydrazine (CCCP), which ablates the mitochondrial membrane potential and causes accumulation of PINK1 on the outer mitochondrial membrane. PINK1 then recruits Parkin from the cytosol, which leads to Parkin-dependent ubiquitination of proteins on the mitochondrial surface. To follow mitophagy in non-neuronal cells, Parkin is often ectopically expressed, as most

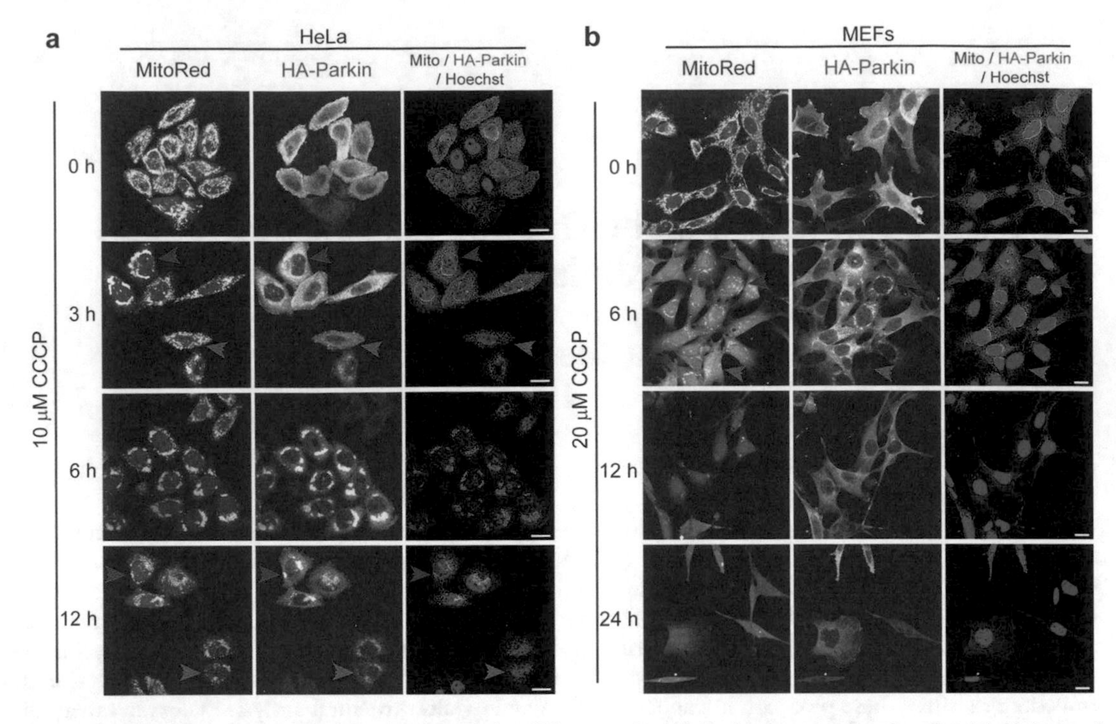

Fig. 1 Immunofluorescence microscopy reveals difference in HA-Parkin expression levels. (**a**) HeLa and (**b**) MEFs were treated with 10 μM (**a**) or 20 μM (**b**) CCCP for the time periods indicated, before immunostaining cells for HA-Parkin (*green*), MitoTracker Red CMXRos (*red*), and Hoechst (*blue*). Our results indicate a variation in HA-Parkin expression levels. In addition HeLa cells and MEFs show different kinetics of mitophagy progression and require different concentrations of CCCP to induce mitophagy. The *arrowheads* show cells expressing high levels (*pink*) or low levels (*blue*) of HA-Parkin. *Scale bars*, 20 μm

tissue culture cell models express very little or no endogenous Parkin. However, the kinetics of mitophagy correlates to Parkin expression levels, which can vary from cell to cell even in a stable cell line. The use of CLEM allows to compare mitophagy progression at a single cell level by ensuring similar levels of Parkin expression (Fig. 1).

For ultrastructural analysis of mitochondria and mitophagy, HA-Parkin wildtype or mutant C431S constructs are transiently expressed in tissue culture cells by retroviral transduction. Since the generation of stable cell lines often results in cells expressing varying levels of Parkin, it is important for cell analysis at the ultrastructural level to select single cells with similar levels of HA-Parkin expression. To compare the amounts of Parkin present, transfected cells expressing HA-Parkin are stained with anti-HA antibody to identify target cells by immunofluorescence microscopy for further analysis by electron microscopy (EM). Light microscopy offers the advantages of rapid screening of a large number of cells, of determining the specificity of labeling of HA-Parkin, and of simultaneously detecting multiple antigens with ease. Electron microscopy, on the other hand, enables the ultrastructure of mitochondria to be determined at high resolution and the different stages of mitophagy to be identified and analyzed. Thus CLEM combines the advantages of both light and EM imaging using the same specimen.

To maintain the fine structure of mitochondria, the first consideration is to choose the optimum buffer for fixation. Traditionally, researchers have used buffers such as phosphate, cacodylate, and HEPES buffers, but these buffers are not entirely innocuous, and some fine structural details may be altered. For example, high concentrations of phosphates may damage mitochondria: therefore, the buffer concentration should be kept as low as possible while still maintaining the pH within the desired range. Several different buffers may be used to protect the cells during fixation. Phosphate and cacodylate buffers predominate in electron microscopy, but organic buffers, such as PIPES [piperazine-1, 4-bis (2-ethanesulfonic acid)], are highly recommended, since it is nontoxic and has less detrimental effects on the cellular fine structure.

CLEM can be used as a combination of fluorescence light microscopy with cryo-EM [3] and cryo-electron tomography [4] or with conventional EM on resin sections. Whereas cryo-EM allows the identification of specific target proteins and detailed analysis of membrane structures, conventional EM on resin sections permits detailed investigations into organelle morphology at high resolution. For analyzing the structure of the mitochondria during the different steps of mitophagy, conventional EM on resin sections is most suitable. Thus, in this chapter, we describe our protocols for correlative light and conventional electron microscopy on resin sections.

2 Materials

2.1 Cell Culture and Transfection

1. MEF (mouse embryonic fibroblast) cells.
2. HEK293 cells.
3. HeLa cells.
4. pMXs-IP HA-Parkin plasmid (Addgene, 38248).
5. Dulbecco's modified Eagle medium (DMEM) + GlutaMAX (Gibco, 31966-021).
6. Modified Eagle medium (MEM) (Sigma, M2279).
7. Roswell Park Memorial Institute-1640 (RPMI-1640) (Sigma, R8758).
8. 10 % fetal bovine serum (FBS) (Sigma, F7524).
9. 2 mM L-glutamine (Sigma, G7513).

10. 100 U/mL penicillin and 10 μg/mL streptomycin (Sigma, P4333).
11. 1 % Nonessential amino acids (Sigma, M7145).
12. No. 1.5 gridded glass bottom 35 mm dish (MatTek, P35G-1.5-14-CGRD).
13. Carbonyl cyanide 3-chlorophenylhydrazone (CCCP) (Acros Organics, AC228131000).
14. Mito Tracker Red CMXRos (Life Technologies, M7512).

2.2 Fixation and Permeabilization of Cultured Cells

1. 16 % (w/v) formaldehyde solution: Methanol free (TAAB Laboratory & Microscopy, F017) (*see* **Note 1**).
2. 0.2 M sodium phosphate buffer, pH 7.4: Prepare 1.0 L of 2× (0.2 M) stock solution with 31.2 g of $NaH_2PO_4{\cdot}2H_2O$ (stock solution A). Prepare 1.0 L stock solution with 28.4 g of Na_2HPO_4 (stock solution B). Autoclave the solutions before storing at room temperature. Prepare a working 0.2 M sodium phosphate buffer solution by mixing 19 mL of solution A and 81 mL of solution B.
3. 0.1 M sodium phosphate buffer, pH 7.4 (PB): Prepare a working 0.1 M PB solution by diluting one part of the 0.2 M sodium phosphate buffer and one part of Milli-Q water and adjust the pH if necessary with NaOH.
4. 4 % (w/v) formaldehyde solution: Mix 3 mL of 16 % (w/v) formaldehyde solution, 6 mL of 0.2 M sodium phosphate buffer, pH 7.4, and 3 mL of Milli-Q water (12 mL total). This solution should be used within 24 h.
5. Milli-Q water produced by a Milli-Q system (Millipore Co., Billerica, MA). Use Milli-Q water in all the solutions.

2.3 Immunostaining

1. Bovine serum albumin (BSA) (Sigma, A2153).
2. Blocking solution: PB containing 3 % (w/v) BSA.
3. Primary antibody: Rat monoclonal antibody for the HA epitope tag (amino acid 98–106 of the human influenza hemagglutinin protein)—clone 3F10, rat IgG (Roche, 11867423001) diluted to 1:100–1:200 with the blocking solution.
4. Secondary antibody: Alexa Fluor 488-conjugated donkey anti-rat IgG [H + L] (Life Technologies, A21208) dilute to 1:300 with blocking solution.
5. Hoechst 33342, 10 mg/mL solution in water, dilute 1:1,000 (Molecular Probes, H-3570).

2.4 Immunofluorescence Light Microscope

1. Confocal microscope: Zeiss LSM 780 Inverted Confocal Microscope using LSM T-PMT point scanning confocal.
2. Lens: ZEISS Plan-APOCHROMAT 40x 1.3 oil DIC.
3. Software: ZEN Black 2012.

2.5 Flat Embedding in Epoxy Resin and Ultramicrotomy

1. 25 % (v/v) glutaraldehyde solution, EM grade (TAAB Laboratory & Microscopy, G011/2) (*see* **Note 1**).
2. PB containing 1 mM glycine.
3. 4 % (w/v) osmium tetroxide (OsO_4) solution (Agar Scientific Limited, R1023) (*see* **Note 1**).
4. 1.5 % (w/v) OsO_4 and 1.5 % (w/v) potassium ferrocyanide (Sigma, 702587) in PB: Dissolve ferrocyanide powder to a final concentration, 1.5 % (w/v), into 1 % (w/v) OsO_4 in PB immediately before use.
5. A graded series of ethanol solutions: 15, 30, 50, 70, 80, 90, and 100 % (v/v) ethanol in Milli-Q water.
6. Agar 100 resin kit (Agar Scientific Limited, R1031): Mix from the resin kit 20 mL of Agar 100 resin, 9 mL of dodecenyl succinic anhydride (DDSA), 12 mL of methyl nadic anhydride (MNA), and 1.2 mL of benzyldimethylamine (BDMA) (*see* **Note 2**).
7. Pre-shaped polyethylene molds with hinged lids; BEEM 3 capsules (Agar Scientific Limited, G362) or BEEM 00 pyramidal capsules (Agar Scientific Limited, G360).

2.6 Sectioning

1. Diamond Knife (Diatome AG).
2. Ultramicrotome (Ultracut UCT, Leica).
3. Single-edge stainless steel razor blades (Fisher Scientific, 11904325).
4. 2 % (v/v) collodion solution (Sigma, 09817).
5. Amyl acetate (Sigma, W504009).
6. 0.5 % (v/v) collodion solution: Mix 5 mL of 2 % (v/v) collodion solution and 15 mL of amyl acetate. This solution should be mixed in fume hood.
7. Thin bar grids (200 mesh, copper) (Agar Scientific Limited). The EM grids are coated with 0.5 % (v/v) collodion solution.
8. Uranyl acetate (Agar Scientific Limited, R1260A).
9. 4 % (v/v) aqueous uranyl acetate solution: Mix 4 g of uranyl acetate and 100 mL of 50 % (v/v) methanol.
10. Sato lead solution (*see* **Note 3**): Weigh out the following, 1.0 g of lead nitrate (Agar Scientific, R1217), 1.0 g of lead acetate (Agar Scientific, R1209), 1.0 g of lead citrate (Agar Scientific,

R1210), 2.0 g of sodium citrate (Sigma, S-1641), 82 mL of boiled Milli-Q water, and 18 mL of 1 N sodium hydroxide (Chemicals VWR BDH Prolab, 1310–73-2). Place compounds in clean Erlenmeyer flask and mix well until the solution becomes clear. Transfer the solution to an amber glass bottle with a screw cap for storage. Before use, filter through 0.22 μm syringe filter (Merck Millipore, SLGP033RS) [5].

3 Methods

3.1 Cell Culture and Transfection

1. HA-Parkin in the pMXs-IP vector was expressed in cells by retroviral transduction.
2. 24 h after transfection the cells were plated on gridded glass bottom dishes.
3. It is important to keep the cells at low density on the glass bottom dishes.
4. After 24 h the cells were grown in the presence or absence of 10 or 20 μM CCCP and 50 nM Mito Tracker Red CMXRos for 3, 6, 12 or 24 h [6].

3.2 Fixation and Permeabilization of Cultured Cells

1. To fix the cells the medium is removed, cells are washed briefly with PB before fixing for at least 2 h or overnight at room temperature with 4 % (w/v) formaldehyde (*see* **Note 4**).
2. After fixing the cells are washed three times for 5 min each with PB.
3. If required the fixed cells in the glass bottom dishes can be kept for several days in PB at 4 °C. The cells should be prevented from drying out.
4. The fixed cells are incubated for 15 s in PB containing 14 % (w/v) glycerol and 35 % sucrose and permeabilized by freezing and thawing in liquid nitrogen for 15 s (*see* **Note 5**).
5. The cells are washed once for 3 min in PB.

3.3 Immunostaining

1. For blocking the cells were incubated for 30 min in PB containing 3 % (w/v) BSA.
2. The blocking solution is replaced by the primary antibody in 3 % BSA in PB.
3. Cells are incubated with the first antibody in a humid chamber overnight at 4 °C.
4. The primary antibody is removed and the cells are washed six times for 10 min with PB.
5. The cells are then incubated in a humid chamber with the Alexa Fluor 488-conjugated second antibody and Hoechst, as the DNA stain, diluted in blocking solution at room temperature for 2 h.

6. The secondary antibody and Hoechst are removed, and the cells are washed six times for 10 min each with PB and kept in PB for imaging.

3.4 Immunofluorescence Light Microscope

1. Map the overall location of cells on the grid by taking DIC images using a 40× lens on the confocal microscope. A series of single slightly overlapping images is taken with the help of a motorized stage in the tile scan mode, which records a defined number of adjoining single images of the whole gridded glass bottom dish.
2. The whole "mosaic" picture is stitched together and the location of the cells identified (Fig. 2a).
3. If immersion oil is used for confocal microscopy, the underside of the glass bottom dish is carefully cleaned with lens cleaner.

3.5 Flat Embedding in Epoxy Resin and Ultramicrotomy

1. Post-fix the cells in PB containing 2.5 % (w/v) glutaraldehyde and 2 % (w/v) formaldehyde on ice for 1.5–2 h (*see* **Note 1**).
2. Wash four times for 10 min in PB containing 1 mM glycine, before rinsing by dipping in a beaker with PB.
3. Post-fix the cells in 1.5 % (w/v) OsO_4 and 1.5 % (w/v) potassium ferrocyanide in PB for 1–2 h on ice in a fume cupboard (*see* **Note 6**).
4. Wash in Milli-Q water three times for 1 min (*see* **Note 7**).
5. Dehydrate the sample on ice in a graded series of ethanol solutions: 15, 30, 50, 70, 80, and 90 % (w/v) ethanol for 10 min each.
6. And then dehydrate the cells at room temperature three times in 100 % (w/v) ethanol for 10 min.
7. Incubate in 50 % (v/v) epoxy resin ethanol mixture for 2 h at room temperature.
8. Then incubate in 75 % (v/v) epoxy resin ethanol mixture for 1–2 h or overnight at room temperature.
9. Finally incubate in 100 % epoxy resin mixture for at least 2 h and up to 12 h at room temperature.
10. A BEEM capsule is filled with 100 % epoxy resin mixture. The BEEM capsule is placed upside-down on the glass bottom dish, and the samples are incubated for 3 days at 60 °C to polymerize the epoxy resin on top of the single layer of cells.
11. After the epoxy resin polymerizes, the epoxy resin block is removed from the glass bottom dish with forceps and plunged briefly into liquid nitrogen.
12. The cells on the newly exposed epoxy resin block surface are observed at 10× magnification under a stereo microscope to identify the region of target cells (Fig. 2a, b). The cells of

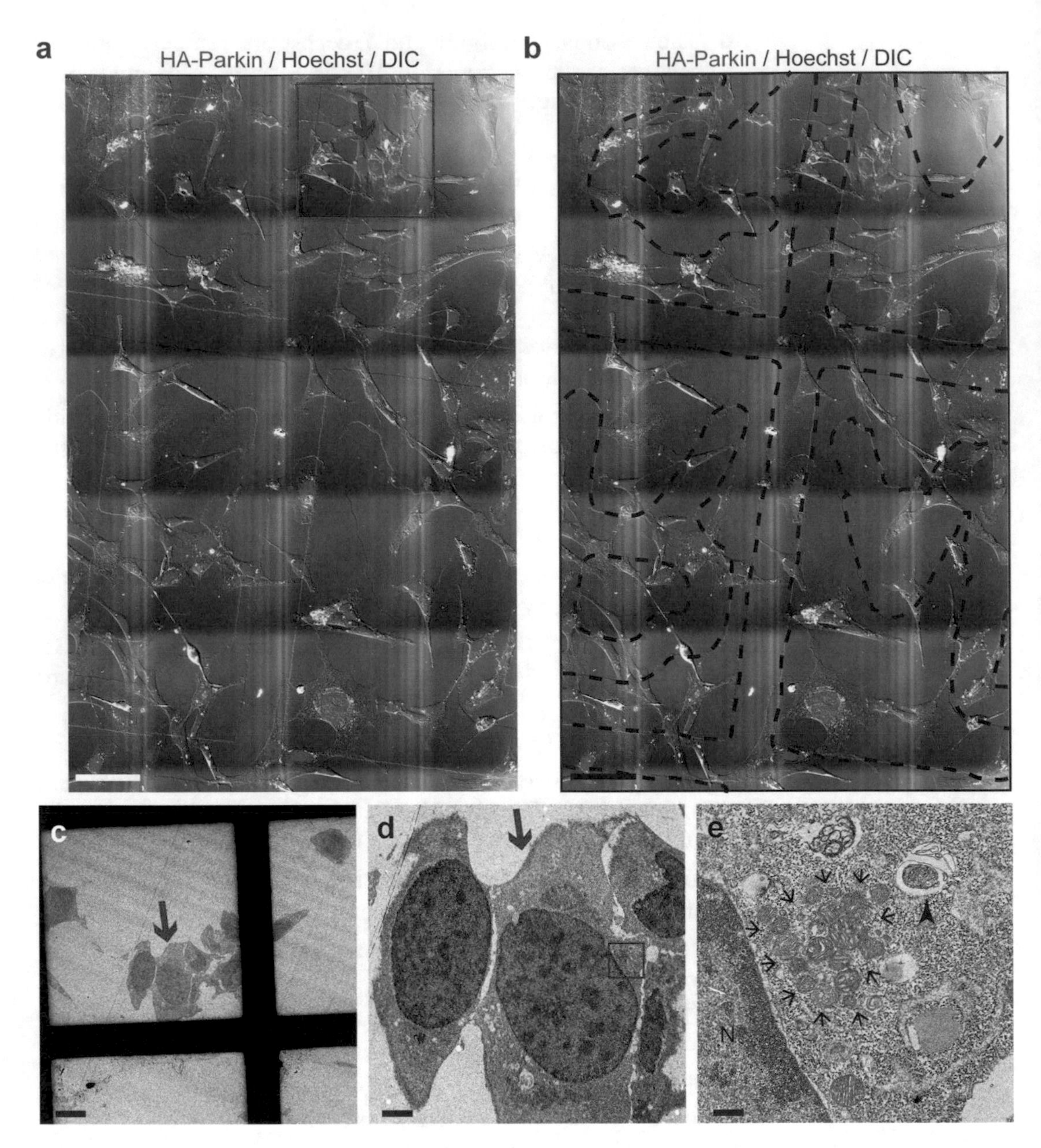

Fig. 2 Example experiment demonstrating the identification of a target cell. (**a**) A cell of interest is first identified on a large differential interference contrast (DIC) and fluorescent microscopic image, which is "stitched" together from 20 single overlapping pictures. (**b**) The letters and lines on the glass bottom dish, marked as *black dotted lines*, are important to find the target cells. (**c**) Low magnification electron micrograph of the square area (*orange*) in panel (**a**). The *orange arrow* highlights the target cell. Panel (**d**) shows the high magnification image of panel (**c**), and panel (**e**) shows the high magnification of panel (**d**). Fragmented mitochondria (*black arrows*) aggregate close to the nucleus (*N*). *Black arrowhead* points at an autophagosome. *Scale bars*, 100 μm (**a**, **b**), 20 μm (**c**), 5 μm (**d**), and 500 nm (**e**)

interest are identified by correlating the grid and cell pattern on the surface of the resin block with the stitched image taken by DIC and fluorescence microscopy.

3.6 Sectioning

1. Trim the block until only a rectangular area of approximately 1 × 2 mm remains that contains the target cells. For the final rectangular cut by hand, use a new razor blade.
2. Fasten the sample block into the sample block holder on the ultramicrotome and align the block face exactly parallel to the diamond knife edge (*see* **Note 8**).
3. Cut thin sections (50–60 nm) right from the start. Although they may not be complete, the very first sections already contain target cells.
4. Collect short ribbons of sections on collodion-coated grids and let them dry.
5. Stain with uranyl acetate for 20 min by placing the grids (section down) onto a 50 μL droplet of 4 % (v/v) aqueous uranyl acetate solution.
6. Wash thoroughly with Milli-Q water.
7. Stain with lead for 5 min by placing the grids on top of a 50 μL droplet of lead staining solution.
8. Wash thoroughly with Milli-Q water and allow to dry on filter paper.

3.7 Electron Microscopy

1. Examine the serial sections in the electron microscope (FEI Tecnai Spirit TEM).
2. Locate the cells and structures of interest identified by fluorescent microscopy in the EM (Fig. 2a, c). Visualize at low magnification the morphology and position of groups or single cells and compare with pictures taken by fluorescent microscopy (*see* Section 3.4).
3. Figure 3 shows representative electron microscope images (Fig. 3) [7].

4 Notes

1. Paraformaldehyde and other chemicals used for fixation, postfixation, and staining are hazardous. Solutions should be prepared and used in a fume hood. Eye protection and gloves should be worn.
2. Epoxy resin mixtures can be stored in syringes sealed with Parafilm for 1 month at −20 °C.

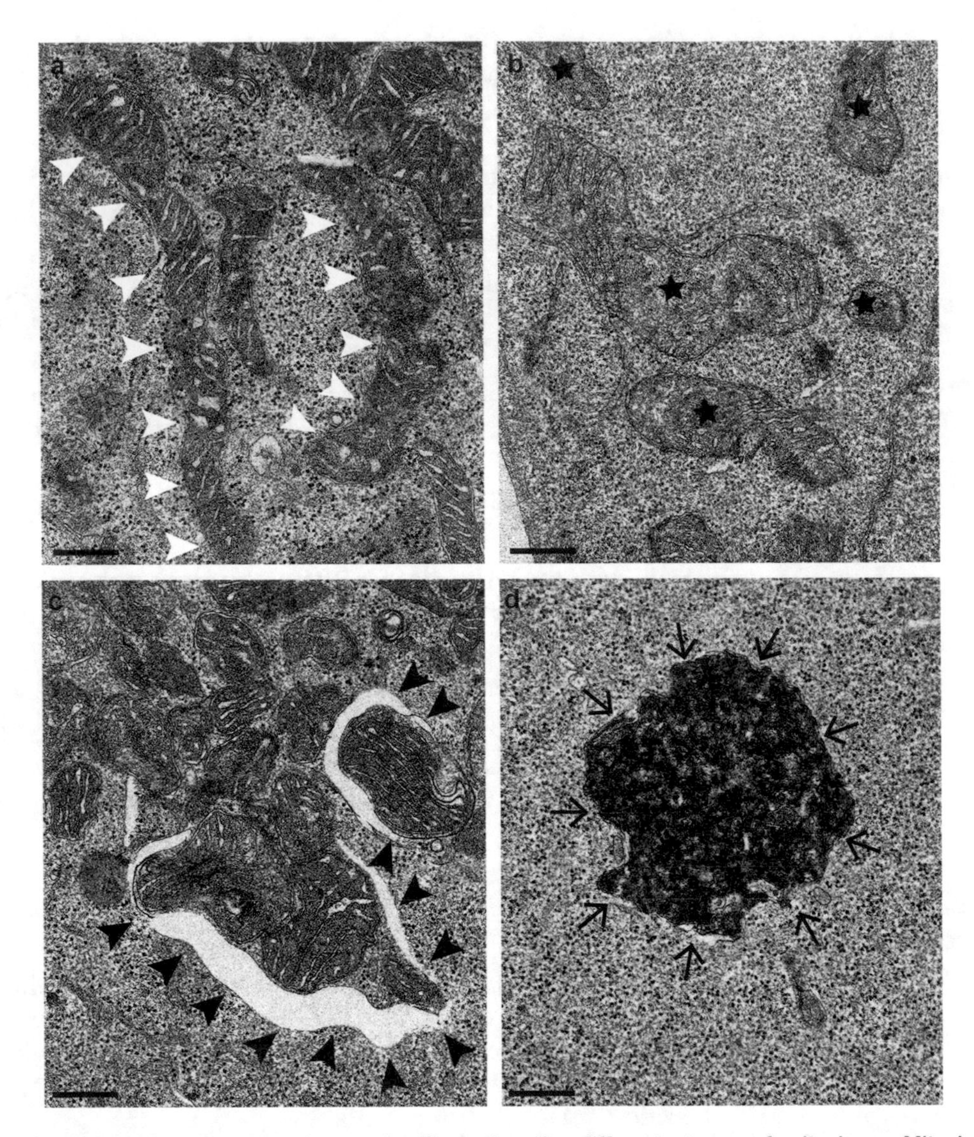

Fig. 3 Representative electron micrographs illustrating the different stages of mitophagy. Mitochondria morphology was analyzed in HA-Parkin expressing HEK293 cells after 0 h (**a**), 2 h (**b**), or 5 h (**c**, **d**) of 10 μM CCCP treatment. In panel (**a**) tubular mitochondria (*white arrow heads*) before CCCP treatment are shown. In (**b**) fragmented mitochondria (*stars*) after 2 h of CCCP treatment are shown. In panel (**c**) after 5 h of CCCP treatment, two phagophores (*black arrowheads*) are shown that surround clusters of fragmented mitochondria. In panel (**d**) an autolysosome is shown (*arrows*) after 5 h of CCCP treatment

3. This lead solution is stable and free from precipitates when kept at room temperature or in the refrigerator for 1 year [5].
4. If the fixative used results in unsatisfactory immunolabeling or ultrastructural preservation, change the composition of the fixative. Adding glutaraldehyde to the paraformaldehyde solution may improve preservation of ultrastructural details. Fix cells in 4 % (w/v) paraformaldehyde containing 0.05–1 % (w/v) glutaraldehyde for 20–30 min.

5. Cells can also be permeabilized by incubating for 30 min in PB containing 0.25 % (w/v) saponin and 5 % BSA. In most cases permeabilization through freeze/thawing in liquid nitrogen leads to better preservation of the ultrastructure as compared to 0.25 % saponin. However, saponin treatment may enhance the permeability of the plasma membrane and access of antibodies to vesicular structures in the cytoplasm.
6. Potassium ferrocyanide enhances the contrast of the membranes and leads to a clearer picture the cellular ultrastructure.
7. The cells can be stored in a refrigerator for a few days before embedding in epoxy resin.
8. It is very important that the block face is properly aligned with the diamond knife. This can be achieved by observing the backlight that passes between the knife and the block face. This light band must have the same width from left to right. Also the distance of the knife must be equal at the top and bottom of the block. This can be achieved by advancing the block at its bottom edge toward the knife until the light band just disappears.

Acknowledgements

We thank Dr. John Kendrick-Jones for his help in the preparation of this chapter, Dr. Antonina J. Kruppa for generating the HEK293 cells stably expressing HA-Parkin, and Dr. Nicholas A. Bright for critical reading of the manuscript. This work was supported by Medical Research Council (MR/K000888/1).

References

1. Kitada T, Asakawa S, Hattori N, Matsumine H, Yamamura Y, Minoshima S, Yokochi M, Mizuno Y, Shimizu N (1998) Mutations in the *parkin* gene cause autosomal recessive juvenile parkinsonism. Nature 392:605–608
2. Narendra D, Tanaka A, Suen DF, Youle RJ (2008) Parkin is recruited selectively to impaired mitochondria and promotes their autophagy. J Cell Biol 183:795–803
3. Sartori A, Gatz R, Beck F, Rigort A, Baumeister W, Plitzko JM (2007) Correlative microscopy: bridging the gap between fluorescence light microscopy and cryo-electron tomography. J Struct Biol 160:135–145
4. Gruska M, Medalia O, Baumeister W, Leis A (2008) Electron tomography of vitreous sections from cultured mammalian cells. J Struct Biol 161:384–392
5. Hanaichi T, Sato T, Malavasi-Yamashiro J, Hoshino M, Mizuno N (1986) A stable lead by modification of Sato's method. J Electron Microsc 35(3):3D4–3D6
6. Itakura E, Kishi-Itakura C, Koyama-Honda I, Mizushima N (2012) Structures containing Atg9A and the ULK1 complex independently target depolarized mitochondria at initial stages of Parkin-mediated mitophagy. J Cell Sci 125:1488–1499
7. Yoshii SR, Kishi C, Ishihara N, Mizushima N (2011) Parkin mediates proteasome-dependent protein degradation and rupture of the outer mitochondrial membrane. J Biol Chem 286:19630–19640

Methods in Molecular Biology (2018) 1759: 41–46
DOI 10.1007/7651_2017_39

Published online: 03 May 2017

Observation of Parkin-Mediated Mitophagy in Pancreatic β-Cells

Atsushi Hoshino and Satoaki Matoba

Abstract

Mitophagy is a cellular process of autophagy-based mitochondrial degradation that eliminates dysfunctional mitochondria and ensures cellular homeostasis. In pancreatic islet β-cells, mitochondria play a pivotal role in glucose-stimulated insulin secretion through ATP production from glucose oxidation. Recent studies have shown that impaired mitophagy and the subsequent mitochondrial compromise contribute to β-cell dysfunction and glucose intolerance. In this chapter, we describe a protocol to monitor Parkin-mediated mitophagy in pancreatic MIN6 β-cells using flow cytometry and a pH-sensitive fluorophore, mKeima.

Keywords Flow cytometry, Mitophagy, mKeima, Pancreatic MIN6 β-cells, Parkin, Retrovirus

1 Introduction

Mitochondrial dysfunction is one of the main mechanisms underlying the progression of type 2 diabetes [1]. Pancreatic β-cells secrete insulin in response to blood glucose via an increase in cytoplasmic ATP from glucose oxidation, leading to cellular depolarization and insulin release [2]. Impaired mitochondria have reduced respiration capacity and high uncoupling proton leak through ROS-mediated UCP2 expression, resulting in a defect in ATP production and subsequent insulin secretion signaling [3]. Moreover, mitochondrial damage increases apoptosis, which underlies the loss of β-cell mass observed in islets from patients with type 2 diabetes [4]. Ultrastructure analysis of diabetic β-cells shows abnormal mitochondria with a rounded shape, distorted cristae, and mitochondrial incorporation into the autophagic vacuole [5, 6], suggesting the importance of mitochondrial quality control for β-cell function.

Mitophagy is an autophagosome-mediated process that sequesters dysfunctional mitochondria. There are several possible mechanisms for the recruitment of autophagosomes to damaged mitochondria. One is the activation of resident mitochondrial proteins that contain an LC3-interacting region (LIR) motif. FUNDC1 and the BH3-only family protein, BNIP3, and NIX are such proteins that are implicated in mitophagy under hypoxic condition and metabolic stress [7, 8]. The other mechanism is the PTEN-induced putative kinase 1 (PINK1)/Parkin system, which utilizes sequestosome-1-like LC3 receptors constituting of p62, optineurin, NBR1, NDP52 and Tax1bp1 [9–11]. PINK1 is a serine/threonine kinase that accumulates in damaged mitochondria via inhibition of proteasome-mediated degradation [12, 13]. Activated PINK1 induces the translocation of Parkin, an E3 ubiquitin ligase, from the cytosol to damaged mitochondria [14], which then mediates the formation of polyubiquitin chains on mitochondrial outer membrane proteins [15]. Sequestosome-1-like LC3 receptors recognize ubiquitinated mitochondria and recruit autophagosomes, followed by lysosomal degradation [16].

Traditional methods to detect mitophagy include electron microscopic observation of mitochondrial remnants within autophagosomes [17] and fluorescent-based co-localization analysis of mitochondrial and autophagosomal markers such as LC3 [18]. However, the utility of electron microscopy for detecting mitophagy is limited by quantification challenges, and LC3-based probes can give false-positive signals due to aggregate formation [19]. Recently, Katayama et al. developed a fluorescent probe, mKeima, to monitor autolysosome maturation. This fluorescent protein, cloned from stony coral, has a bimodal excitation spectrum with peaks 440 and 586 nm, which corresponds to the neutral and acidic conditions of cellular compartments [20]. Use of the mitochondrial-targeted form of mKeima (mt-mKeima) allows one to determine whether mitochondria are in the lysosome (pH ~4.5) or in the cytoplasm (pH ~7.5). Using the example of a Parkin-expressing mouse pancreatic β-cell line MIN6 cells [21], this chapter describes a flow cytometry-based method to detect mitophagy with mt-mKeima.

2 Materials

2.1 Cell Culture and Viral Gene Transfer

1. MIN6 cells: mouse pancreatic β-cell line, propagated in DMEM with GlutaMAX™ supplement, containing 10% fetal bovine serum (FBS), 50 μM β-mercaptoethanol, 100 U/100 μg/mL penicillin/streptomycin at 37 °C with 5% CO_2.
2. HEK293T cells: packaging cell line, propagated in DMEM with GlutaMAX™ supplement, containing 10% FBS at 37 °C with 5% CO_2.
3. Fugene HD transfection reagent (Promega).
4. 28 mm Diameter Syringe Filters, 0.45 μm Pore SFCA Membrane, Sterile.
5. Polybrene (Hexadimethrine bromide; Sigma-Aldrich).

6. Antibiotics: Geneticin Selective antibiotic (G418 Sulfate; Thermo Fisher Scientific).
7. Carbonyl cyanide m-chlorophenylhydrazone (CCCP; Sigma-Aldrich); dissolved in DMSO. Store at −20 °C.
8. Trypsin 2.5%.

2.2 Plasmid

1. pMSCV mt-mKeima-P2A-hParkin: mt-mKeima and P2A linked human Parkin, cloned into pMSCV, a Retrovirus transfer plasmid (*see* **Note 1**).
2. Gag/pol (addgene #14887): Retroviral packaging plasmid.
3. VSV.G (addgene #14888): Retroviral envelope plasmid.

2.3 Flow Cytometry

1. Assay buffer: Dulbecco's phosphate-buffered saline (PBS) containing 2% FBS and 20 mM HEPES.
2. Flow cytometry tube: 5 mL Round Bottom Polystyrene Test Tube (Corning).
3. BD™ LSR II flow cytometer.

3 Methods

3.1 Retrovirus Generation

1. Seed ~5 × 10^6 HEK293T cells on a 10 cm plate and incubate overnight until the cells reach ~80% confluence.
2. Replace with 10 mL of fresh media 2 h before transfection.
3. Mix retroviral plasmids (10 μg transfer plasmid, 10 μg gal/pol, 5 μg VSV.G) in 600 μL of complete media in an eppendorf tube. Add 50 μL of Fugene HD directly to the DNA mixture. Gently vortex tube containing transfection mixture and incubate at room temperature for 10 min (*see* **Note 2**).
4. Add transfection mixture dropwise to the cells, incubate 4 h to overnight, and replace with fresh media.
5. Collect virus-containing media 48 h after transfection (*see* **Note 3**).
6. Pass viral media through a 0.45 μm low protein-binding filter. At this point, the viral supernatant can be used to infect cells or frozen at −80 °C.

3.2 Retrovirus Transduction

1. Seed ~1 × 10^6 MIN6 cells per well were plated into a 6-well plate and incubate overnight.
2. Replace with 2 mL virus media and 2 mL complete media and add 8 μg/mL polybrene (final concentration).
3. Centrifuge the 6-well plate at 1,000 × *g* for 90 min at 37 °C (*see* **Note 4**).
4. After the spin, change media to decrease polybrene toxicity.

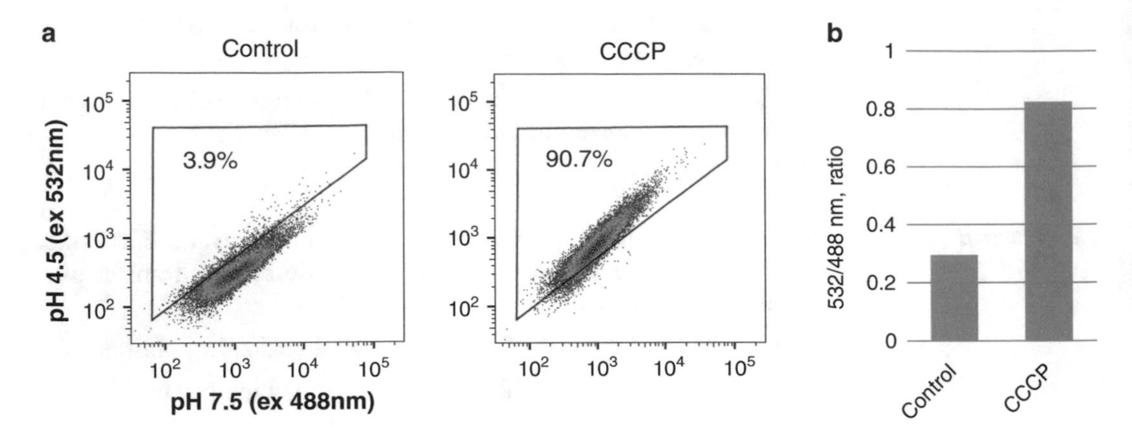

Fig. 1 CCCP treatment as positive control for mitophagy induction in MIN6 cells. Parkin-expressing cells were treated with 10 μM CCCP for 24 h. Representative scatter image of flow cytometry for mt-mKeima (**a**) and the ratio of mean value of 532/488 nm (**b**) for control and CCCP-treated cells

5. 24 h after infection, treat with the appropriate selection drug, 500 μg/mL G418 for 7 days (*see* **Note 5**).

3.3 Mitophagy Induction as a Positive Control

To induce mitophagy in Parkin-expressing MIN 6 cells, treat with 10 μM CCCP in complete media for 24 h.

3.4 Flow Cytometry

1. Prepare samples in a 12-well plate.
2. Wash with PBS twice and add 0.2 mL 2.5% Trypsin, followed by 5 min incubation at 37 °C.
3. Neutralize with 1 mL assay buffer and transfer cell suspensions to flow cytometry tube.
4. Centrifuge the tube at 500 × *g* for 5 min at 4 °C.
5. Aspirate the supernatants completely and resuspend cell pellets in 0.3 mL assay buffer.
6. Keep the samples on ice and in the dark.
7. Measure the samples in LSR II flow cytometer with the following laser and filter setting; ex 488 nm–em 695/40 nm and ex 532 nm–em 610/20 nm (*see* **Note 6**).
8. Analyze the data using FlowJo software (Fig. 1).

4 Notes

1. Mt-mKeima plasmid is available in MBL and Addgene (#56018, # 72342), and human Parkin template plasmid is available in Addgene (#23956, etc.).
2. Other transfection agent and calcium phosphate method can be used as well.

3. Virus media can be collected repeatedly every 12 h.
4. Plate centrifuge (spinfection) increases the transduction efficiency.
5. This step is not essential. Usually this protocol setting achieves high transduction efficiency and also mKeima positive cell can be gated in the step of data analysis.
6. Instead of yellow laser, green (532 nm) laser is available to monitor acidic pH signal (586 nm) of mKeima.

Acknowledgements

This study was supported in part by Grants-in-Aid from the Ministry of Education, Science and Culture of Japan (MEXT) (KAKENHI 15K09144 to SM).

References

1. Mulder H, Ling C (2009) Mitochondrial dysfunction in pancreatic beta-cells in type 2 diabetes. Mol Cell Endocrinol 297(1–2):34–40
2. Fajans SS, Bell GI, Polonsky KS (2001) Molecular mechanisms and clinical pathophysiology of maturity-onset diabetes of the young. N Engl J Med 345(13):971–980
3. Zhang CY et al (2001) Uncoupling protein-2 negatively regulates insulin secretion and is a major link between obesity, beta cell dysfunction, and type 2 diabetes. Cell 105(6):745–755
4. Szabadkai G, Duchen MR (2009) Mitochondria mediated cell death in diabetes. Apoptosis 14(12):1405–1423
5. Lu H, Koshkin V, Allister EM, Gyulkhandanyan AV, Wheeler MB (2010) Molecular and metabolic evidence for mitochondrial defects associated with beta-cell dysfunction in a mouse model of type 2 diabetes. Diabetes 59 (2):448–459
6. Hoshino A et al (2014) Inhibition of p53 preserves Parkin-mediated mitophagy and pancreatic beta-cell function in diabetes. Proc Natl Acad Sci U S A 111(8):3116–3121
7. Liu L et al (2012) Mitochondrial outer-membrane protein FUNDC1 mediates hypoxia-induced mitophagy in mammalian cells. Nat Cell Biol 14(2):177–185
8. Zhang J, Ney PA (2009) Role of BNIP3 and NIX in cell death, autophagy, and mitophagy. Cell Death Differ 16(7):939–946
9. Heo JM, Ordureau A, Paulo JA, Rinehart J, Harper JW (2015) The PINK1-PARKIN mitochondrial ubiquitylation pathway drives a program of OPTN/NDP52 recruitment and TBK1 activation to promote mitophagy. Mol Cell 60(1):7–20
10. Lazarou M et al (2015) The ubiquitin kinase PINK1 recruits autophagy receptors to induce mitophagy. Nature 524(7565):309–314
11. Wong YC, Holzbaur EL (2014) Optineurin is an autophagy receptor for damaged mitochondria in Parkin-mediated mitophagy that is disrupted by an ALS-linked mutation. Proc Natl Acad Sci U S A 111(42):E4439–E4448
12. Takatori S, Ito G, Iwatsubo T (2008) Cytoplasmic localization and proteasomal degradation of N-terminally cleaved form of PINK1. Neurosci Lett 430(1):13–17
13. Narendra DP et al (2010) PINK1 is selectively stabilized on impaired mitochondria to activate Parkin. PLoS Biol 8(1):e1000298
14. Vives-Bauza C et al (2010) PINK1-dependent recruitment of Parkin to mitochondria in mitophagy. Proc Natl Acad Sci U S A 107 (1):378–383
15. Geisler S et al (2010) PINK1/Parkin-mediated mitophagy is dependent on VDAC1 and p62/SQSTM1. Nat Cell Biol 12(2):119–131
16. Birgisdottir AB, Lamark T, Johansen T (2013) The LIR motif—crucial for selective autophagy. J Cell Sci 126(Pt 15):3237–3247

17. Mizushima N, Levine B (2010) Autophagy in mammalian development and differentiation. Nat Cell Biol 12(9):823–830
18. Dolman NJ, Chambers KM, Mandavilli B, Batchelor RH, Janes MS (2013) Tools and techniques to measure mitophagy using fluorescence microscopy. Autophagy 9 (11):1653–1662
19. Kuma A, Matsui M, Mizushima N (2007) LC3, an autophagosome marker, can be incorporated into protein aggregates independent of autophagy: caution in the interpretation of LC3 localization. Autophagy 3(4):323–328
20. Katayama H, Kogure T, Mizushima N, Yoshimori T, Miyawaki A (2011) A sensitive and quantitative technique for detecting autophagic events based on lysosomal delivery. Chem Biol 18(8):1042–1052
21. Miyazaki J et al (1990) Establishment of a pancreatic beta cell line that retains glucose-inducible insulin secretion: special reference to expression of glucose transporter isoforms. Endocrinology 127(1):126–132

Methods in Molecular Biology (2018) 1759: 47–57
DOI 10.1007/7651_2017_9

Published online: 22 March 2017

Monitoring Mitochondrial Changes by Alteration of the PINK1-Parkin Signaling in *Drosophila*

Tsuyoshi Inoshita, Kahori Shiba-Fukushima, Hongrui Meng, Nobutaka Hattori, and Yuzuru Imai

Abstract

Mitochondrial quality control is a key process in tissues with high energy demands, such as the brain and muscles. Recent studies using *Drosophila* have revealed that the genes responsible for familial forms of juvenile Parkinson's disease (PD), *PINK1* and *Parkin* regulate mitochondrial function and motility. Cell biological analysis using mammalian cultured cells suggests that the dysregulation of mitophagy by PINK1 and Parkin leads to neurodegeneration in PD. In this chapter, we describe the methods to monitor mitochondrial morphology in the indirect flight muscles of adult *Drosophila* and *Drosophila* primary cultured neurons and the methods to analyze the motility of mitochondria in the axonal transport of living larval motor neurons.

Keywords: Axonal transport, Fluorescence imaging, Mitochondrial fusion and fission, Muscle mitochondria, Primary neuron culture

1 Introduction

The *parkin* gene, which encodes a ubiquitin ligase (E3) with properties of both HECT-type and RING finger-type E3 ligases, is known to be responsible for an autosomal recessive form of early-onset familial PD [1]. The *PINK1* gene, which encodes a mitochondrial kinase, is another gene responsible for an autosomal recessive form of early-onset familial PD [2]. A series of cell biological studies has demonstrated that PINK1 and Parkin are involved in mitophagy, a mitochondrial quality control response [3–5]. PINK1 is required for the recruitment of Parkin to damaged mitochondria, where Parkin induces mitophagy through its E3 activity [6–11].

Drosophila is a powerful animal model to determine the roles of PINK/Parkin signaling in vivo. Loss of the *PINK1* or *Parkin* genes causes remarkable degeneration of mitochondria in the indirect flight muscles and sperm of adult flies, which allows genetic screening to isolate genes involved in this signaling [12–17]. In fact, *mitofusin* (*Mfn*) and *miro* have been isolated as genes that modulate the mitochondrial phenotypes of *PINK1/Parkin* mutant flies and have been characterized as Parkin ubiquitination substrates during mitophagy [18–21]. Suppression of mitochondrial motility by Miro degradation is important to retain damaged mitochondria in the cell bodies in neurons, whereas fragmentation of mitochondria via Mfn degradation facilitates the effective removal of damaged mitochondria by mitophagy. These events can be analyzed in larval motor neurons and cultured neurons prepared from embryos [20–23]. The observation of mitochondria in the indirect flight muscles is another useful assay because they exhibit a highly homogeneous configuration in their morphology [16, 17]. In this chapter, we will introduce methods to analyze mitochondrial morphology and motility in *Drosophila* cultured cells and tissues.

2 Materials

2.1 Visualization of Mitochondria in the Muscle Tissues

1. PBS-T: Phosphate-buffered saline (PBS) containing 0.3% Triton X-100.
2. Fixative solution: 4% paraformaldehyde dissolved in PBS (store at 4°C).
3. Tetramethylrhodamine (TRITC)-labeled phalloidin (Sigma-Aldrich, P1951) stock solution: 1 μg/ml dissolved in methanol (store at −20°C).
4. TRITC-labeled phalloidin staining solution: TRITC-labeled phalloidin stock solution diluted at 1:40 with PBS-T.
5. Mounting agent: SlowFade Gold Antifade Mountant (Thermo Fisher Scientific, S36936) or equivalent.
6. Colorless nail polish.
7. *Drosophila melanogaster* harboring gene mutations or transgenes (most strains are available from public stock centers that can be accessed through the web site http://flybase.org/wiki/FlyBase:Stocks).
8. Transgenic *Drosophila* line P{UAS-mito-HA-GFP.AP}2, also known as UAS-mitoGFP (Bloomington *Drosophila* Stock Center, Stock Nos. 8442 and 8443).
9. Tweezers (FST, Item#: 11251-30, Dumont #5 forceps Dumoxel standard tip).
10. Microscissors (WPI, Item#: 501778, SuperFine Vannas scissors, 3 mm straight blades).
11. 1.5 ml Eppendorf-type tube.
12. 96-well plate (round or flat bottom).
13. Black or dark plastic tape (200 μm thick).

2.2 Live Imaging of Axonal Transport of Mitochondria

1. HL-3 solution: 70 mM NaCl, 5 mM KCl, 20 mM $MgCl_2$, 10 mM $NaHCO_3$, 115 mM sucrose, 5 mM trehalose, 5 mM HEPES (pH 7.2), and 2 mM $CaCl_2$. Filter the solution with a syringe filter unit (Millipore, Item#: SLGS033SB, Millex-GS, 0.22 μm) and store at 4°C for 2 weeks.
2. Silicone (Shin-Etsu silicone, KE-106 and CAT-RG).
3. *Drosophila melanogaster* harboring gene mutations or transgenes.
4. UAS-mitoGFP lines.
5. Tweezers (FST, Item#: 11251-30, Dumont #5 forceps Dumoxel standard tip).
6. Microscissors (WPI, Item#: 501778, SuperFine Vannas scissors, 3 mm straight blades).
7. 35-mm culture dishes.
8. Insect pins (FST, Minutien pins, tip diameter 0.0125 mm, rod diameter 0.1 mm, stainless steel). Cut the heads short (to ~3 mm in length) so as not to scratch the objective lens.

2.3 Mitochondrial Imaging of Primary Cultured Neurons Derived from Neuroblasts

1. Grape juice agar plate (*see* **Note 1**): Autoclave 25–30 g of agar with 700 ml H_2O for 40 min (solution A). Add 20 ml of 95% ethanol to a tube containing 0.5 g of methyl 4-hydroxybenzoate (Sigma-Aldrich, H3647). Add the solution of methyl 4-hydroxybenzoate to 300 ml of 100% grape juice concentrate (solution B). Add solution B to solution A and pour the mixture into 10-cm plastic plates (*see* **Note 2**).
2. Embryo washing buffer: 0.7% NaCl, 0.05% Triton X-100, autoclaved.
3. Sterilized water: Autoclaved Milli-Q water.
4. Schneider's *Drosophila* medium (GIBCO, 21720024).
5. Penicillin–streptomycin solution (×100) (GIBCO, 15140122).
6. Fetal bovine serum for insect cell culture (Sigma-Aldrich, F0643).
7. 10- or 6-cm plastic dishes (petri dishes).
8. Sterilized 1.5 ml Eppendorf-type tubes.
9. Embryo-collecting container: Multiples required for multiple genotypes.
10. Dounce tissue homogenizer with a loose pestle: Working volume 1 ml. Multiples required for multiple genotypes.
11. 40-μm cell strainer (Falcon, 352340).
12. 35-mm glass bottom dish (MatTek, poly-D-lysine coated, P35GC-0-10-C) or multi-well chamber (Nunc, Lab-Tek II Chamber Slide).

3 Methods

3.1 Visualization of Mitochondria in the Muscle Tissues

1. Anesthetize adult fly crosses expressing mitoGFP in the muscle tissues with carbon dioxide.
2. Cut off the abdomens and then cut off the heads with microscissors (Fig. 1a, *see* **Note 3**).
3. Pick up the thoraxes anchoring wings or legs by tweezers and rinse thoraxes briefly with PBS-T so as not to shed the fixative solution and sink the thoraxes in an Eppendorf tube containing 1 ml of fixative solution (>6 thoraxes/tube).

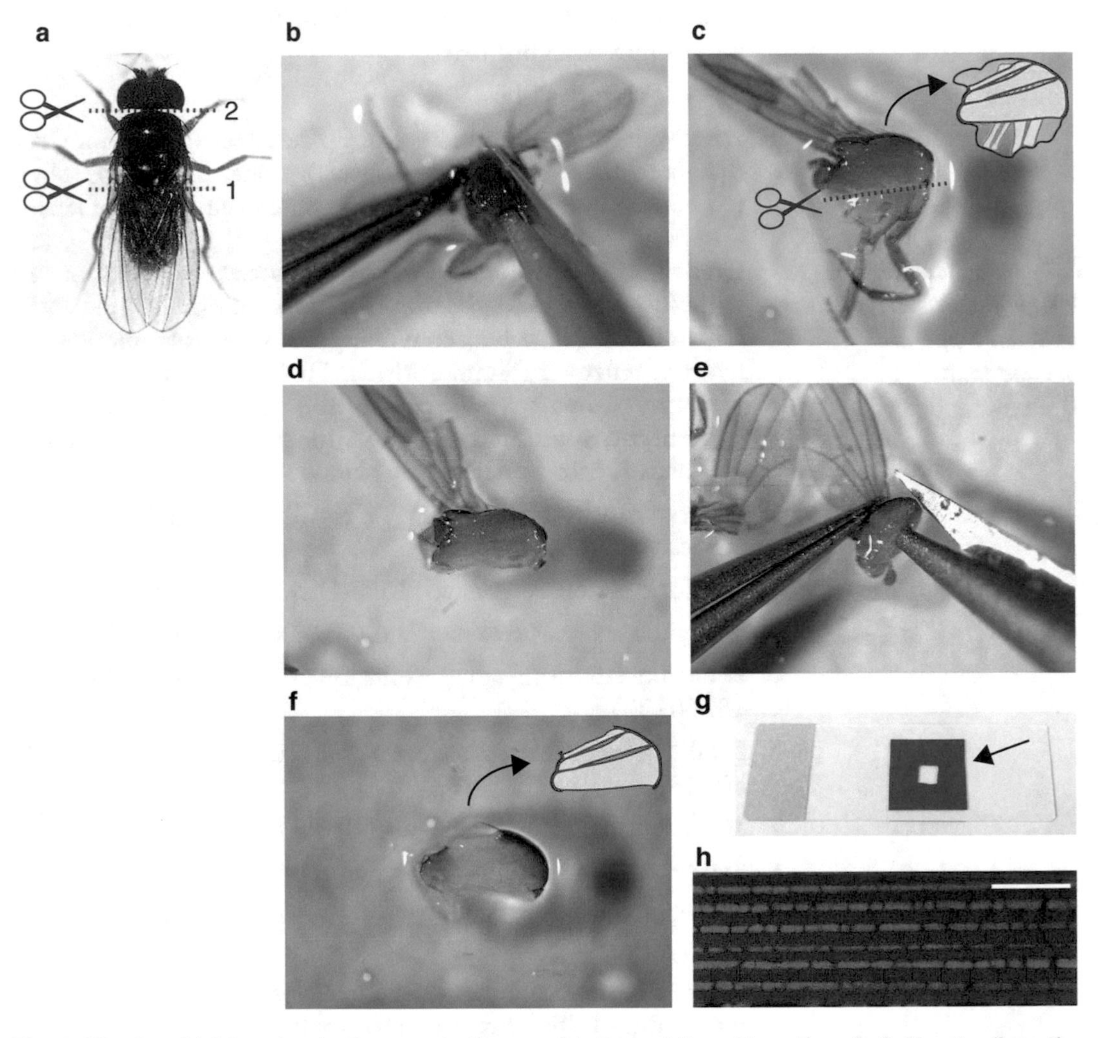

Fig. 1 Mitochondrial imaging in the muscle tissues. (**a**) *Drosophila* cutting sites. (**b-f**) Muscle dissection procedure. A *dashed line* indicates the cutting position. The dorsal longitudinal muscles and the dorsoventral muscles are depicted in *beige* and *green*, respectively, in **c** and **f**. (**g**) Plastic tape window (*arrow*) on a microscope slide. (**h**) High-magnification image of muscular mitochondria visualized with mitoGFP (*green*), counterstaining myofibers with TRITC-phalloidin (*red*). Bar = 10 μm

4. Rotate the tube overnight at 4°C (or 2 h at RT).
5. Wash the thoraxes with 1 ml PBS twice.
6. Isolate the indirect flight muscles using microscissors and tweezers following the below steps: Incise the dorsal side of the thorax along with the midline with microscissors (Fig. 1b, *see* **Note 4**). Incise the ventral side in a similar way and divide the thorax in two. Cut off the legs with microscissors (Fig. 1c, d). Incise the border between the dorsal longitudinal muscles (depicted in beige, Fig. 1c) and dorsoventral muscles (depicted in green, Fig. 1c). Harvest the dorsal longitudinal muscles (Fig. 1f).
7. Stain with 100 μl TRITC-labeled phalloidin staining solution/well on a 96-well plate overnight at 4°C.
8. Make a window (~5 mm squares) in the center of a piece of plastic tape that is slightly larger than a cover glass on a microscope slide and attach the tape to the microscope slide (Fig. 1g).
9. Transfer the muscle fascicles onto the tape window of a microscope slide using a 200-μl pipette tip whose tip has been cut off with scissors.
10. Wash the muscle fascicles with 20 ~ 30 μl PBS twice on the microscope slide.
11. Add a drop of mounting agent, cover the tissues with a cover glass and seal the cover glass with colorless nail polish.
12. Take images using a confocal microscope (Fig. 1h, *see* **Note 5**).

3.2 Live Imaging of Axonal Transport of Mitochondria

1. To prepare a dissection dish, mix 3.6 ml KE-106 and 0.4 ml CAT-RG in a 35-mm culture dish and incubate them at 37°C for 2 days to congeal the silicone (Fig. 2a).
2. Dissect larvae expressing mitoGFP in motor neurons on the dissection dish following the steps described below: Pick up the third instar larva and wash the body surface in HL-3 solution. Transfer the larva into 2 ~ 3 ml of HL-3 solution on the dissection dish. Hold both the front and tail edges of the larva with insect pins (red arrowheads in Fig. 2b), keeping its dorsal side upward so that tracheae can be seen through the body wall. Cut the body wall along the dorsal midline (dashed line in Fig. 2b) to make a fillet, holding four corners of the body wall with insect pins (blue arrowheads in Fig. 2b). Remove the gut and salivary glands carefully, refraining from injuring the ventral ganglion (Fig. 2b, c), axons (Fig. 2c), and muscles (*see* **Note 6**).
3. Replace the solution with 2 ~ 3 ml of fresh HL-3 solution and observe a motor neuron axon (Fig. 2c) under an upright microscope. Take live images every 1 s for 60 s with a high-power

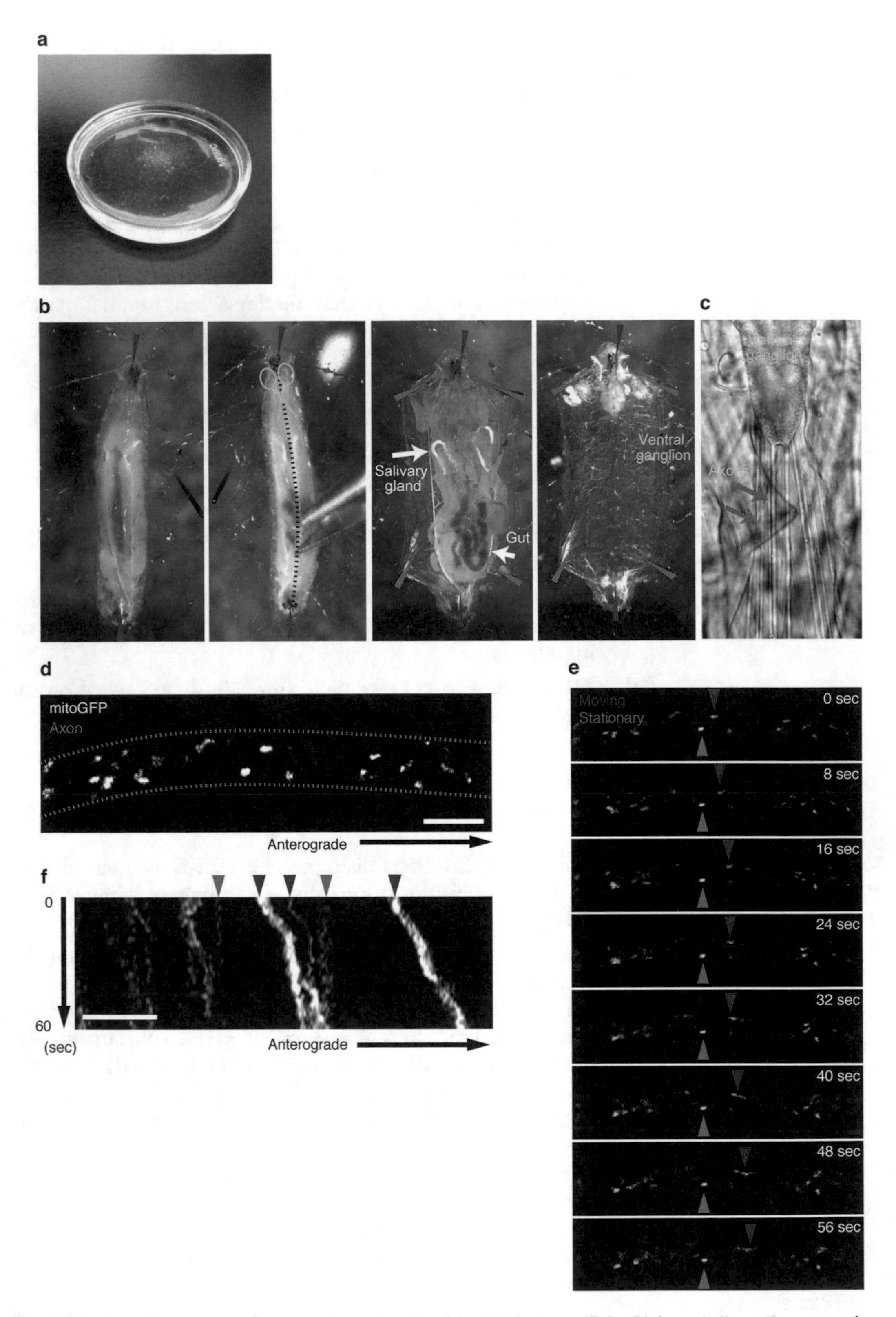

Fig. 2 Live imaging of axonal transport of mitochondria. (**a**) Silicone dish. (**b**) Larval dissection procedure. *Arrowheads* indicate the positions of insect pins. A *black dashed line* indicates cutting position. (**c**) Ventral

(×60–100) water immersion objective (Fig. 2d, e). Make a kymograph (Fig. 2f) using ImageJ (select regions of interest [ROIs] of images and make a kymograph using Image > Stacks > Reslice menu).

3.3 Mitochondrial Imaging of Primary Cultured Neurons Derived from Neuroblasts

1. Set up a fly cross in a fly embryo-collecting container on a grape juice agar plate with a yeast chunk on the center (Fig. 3a–c, *see* **Note 7**).
2. Replace the grape juice agar plate with a new one without a yeast chunk by turning the container upside down and mildly anesthetizing the flies with CO_2. Incubate at 25°C for 2 h to collect embryos (*see* **Note 8**).
3. Remove the agar plate and further incubate at 25°C for 4 h (*see* **Note 9**).
4. Prepare Schneider's *Drosophila* medium containing 5% fetal bovine serum (FBS) and 1× penicillin–streptomycin (culture medium) and keep it at 4°C. Prepare bleach solution (40% bleach in water) just before use. Chill an autoclaved Dounce tissue homogenizer with a loose pestle on ice.
5. Remove the plate from the 25°C incubator and rinse the plate with the embryo washing buffer. Collect embryos from the plate with a cotton tip, and transfer them with the embryo washing buffer into a sterilized Eppendorf tube with a 1-ml plastic pipette.
6. Remove the embryo washing buffer after the embryos settle down, and wash them with the washing buffer twice and subsequently with sterilized water twice.
7. Remove water after embryos settle down.
8. Dechorionate the embryos with 1 ml of 40% bleach for 2 min (*see* **Note 10**).
9. Remove the bleach immediately.
10. Rinse the embryos with 1 ml of culture medium four times (*see* **Note 11**).
11. Resuspend the embryos with 1 ml of culture medium, and transfer into an autoclaved Dounce tissue homogenizer.

Fig. 2 (Continued) ganglion and axons of motor neurons (*arrows*). (**d**) High-magnification image of mitochondria visualized with mitoGFP in a motor neuron axon (*yellow lines*). Bar = 5 μm. (**e**) Time-lapse images of mitoGFP as shown in **d**. Some of the mitochondria move to the nerve terminal (*red arrowhead*), while others are stationary. (**f**) Kymograph tracking mitochondrial movement over 60 s. *Blue and red arrowheads* indicate stationary and moving mitochondria, respectively. Bar = 5 μm

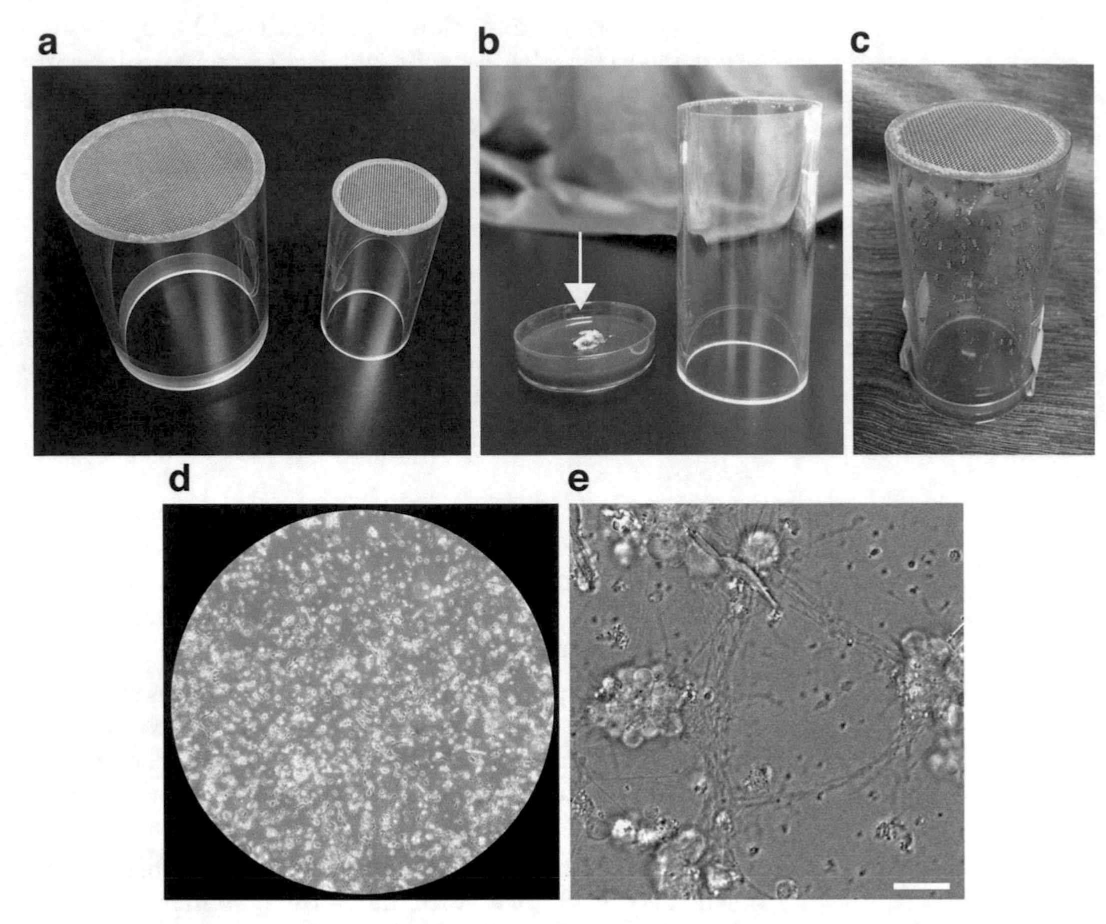

Fig. 3 Primary cultured neurons derived from neuroblasts. (**a**) Embryo-collecting containers for 10-cm (*left*) and 6-cm dishes (*right*). The 10-cm dish type is used for over 100 flies, and the 6-cm dish type is for ~100 flies. (**b**) A grape juice agar plate with a yeast chunk in the center (*arrow*) and a 6-cm dish-type container (*right*). (**c**) Assembled embryo-collecting container with fly crosses. (**d**) Phase-contrast micrograph image of primary culture 24 h post-plating (low-magnification image). (**e**) Phase-contrast micrograph image of primary culture 72 h post-plating

12. Move the Dounce pestle slowly up and down 10–12 times.
13. Filter the homogenized embryos with a 40-μm cell strainer and a sterilized 50-ml Falcon conical tube. Preparation should be performed on a clean bench after this step.
14. Transfer the filtered homogenate into a sterilized Eppendorf tube, and spin down neuroblasts at 1,000 *g* for 3 min at 4°C.
15. Remove the supernatant carefully, resuspend the cell pellet in 1.0 ml culture medium (*see* **Note 12**), and spin down at 1,000 × *g* for 3 min at 4°C.
16. Repeat **step 15** two or three more times.
17. Add 500 μl of culture medium to the cell pellet and gently pipette 60 times to make a single-cell suspension.

18. Count the cell number, and plate the cells on a culture dish or in a chamber at 50–70% confluence (1.0×10^5 cells/200 μl) (Fig. 3d, *see* **Note 13**).
19. Incubate in a humidified 25°C chamber and carefully exchange half of the culture medium to remove debris 24 h after plating (*see* **Note 14**).

4 Notes

1. An extra number of plates should be prepared to increase the odds of observing a peak of egg-laying.
2. Perform the procedure quickly to avoid the solidification of the mixture in the pouring container.
3. Leave legs and wings to facilitate the transfer of the thorax.
4. Insert a blade to the base of the neck in the thorax, and cut the dorsal side of the thorax in one stroke. Repeated incisions damage muscle fascicles.
5. Due to the thickness of tissues, the TRITC staining might be weaker in the middle of tissues. Focusing on the tissue surface will result in better images.
6. To maintain the mitochondria in good condition, it is important to dissect the larva within 5 min.
7. Use young flies (3 ~ 7 days post-eclosion) to obtain enough eggs; a peak of egg-laying will occur approximately 5–7 days after mating. Mating can be achieved in either a fly bottle or the embryo-collecting container. When the UAS-mitoGFP fly line is used, the mitochondria can be visualized with GFP fluorescence.
8. Approximately 5×10^4 cells can be obtained from 100 embryos.
9. Incubate until neuroblasts emerge.
10. Longer incubation reduces cell viability.
11. Centrifugation is not required. Embryos precipitate quickly.
12. Two tubes containing the culture medium should be prepared. One is to use when away from a clean bench, and the other is for use at a clean bench.
13. When a 35-mm glass bottom dish is used, plate 300–500 μl of the cell suspension on the glass bottom region. After the cells attach to the glass, add 2 ml of the culture medium slowly from the edge of the dish.
14. Neurites emerge 1 day after plating, and typical neuronal morphology can be observed by 3–7 days (Fig. 3e).

References

1. Kitada T, Asakawa S, Hattori N, Matsumine H, Yamamura Y, Minoshima S, Yokochi M, Mizuno Y, Shimizu N (1998) Mutations in the parkin gene cause autosomal recessive juvenile parkinsonism. Nature 392 (6676):605–608. doi:10.1038/33416
2. Valente EM, Abou-Sleiman PM, Caputo V, Muqit MM, Harvey K, Gispert S, Ali Z, Del Turco D, Bentivoglio AR, Healy DG, Albanese A, Nussbaum R, Gonzalez-Maldonado R, Deller T, Salvi S, Cortelli P, Gilks WP, Latchman DS, Harvey RJ, Dallapiccola B, Auburger G, Wood NW (2004) Hereditary early-onset Parkinson's disease caused by mutations in PINK1. Science 304(5674):1158–1160. doi:10.1126/science.1096284. [pii]
3. Narendra D, Tanaka A, Suen DF, Youle RJ (2008) Parkin is recruited selectively to impaired mitochondria and promotes their autophagy. J Cell Biol 183(5):795–803. doi:10.1083/jcb.200809125
4. Matsuda N, Sato S, Shiba K, Okatsu K, Saisho K, Gautier CA, Sou YS, Saiki S, Kawajiri S, Sato F, Kimura M, Komatsu M, Hattori N, Tanaka K (2010) PINK1 stabilized by mitochondrial depolarization recruits Parkin to damaged mitochondria and activates latent Parkin for mitophagy. J Cell Biol 189(2):211–221. doi:10.1083/jcb.200910140
5. Geisler S, Holmstrom KM, Skujat D, Fiesel FC, Rothfuss OC, Kahle PJ, Springer W (2010) PINK1/Parkin-mediated mitophagy is dependent on VDAC1 and p62/SQSTM1. Nat Cell Biol 12(2):119–131. doi:10.1038/ncb2012
6. Shiba-Fukushima K, Imai Y, Yoshida S, Ishihama Y, Kanao T, Sato S, Hattori N (2012) PINK1-mediated phosphorylation of the Parkin ubiquitin-like domain primes mitochondrial translocation of Parkin and regulates mitophagy. Sci Rep 2:1002. doi:10.1038/srep01002
7. Kane LA, Lazarou M, Fogel AI, Li Y, Yamano K, Sarraf SA, Banerjee S, Youle RJ (2014) PINK1 phosphorylates ubiquitin to activate Parkin E3 ubiquitin ligase activity. J Cell Biol 205(2):143–153. doi:10.1083/jcb.201402104
8. Koyano K, Okatsu K, Kosako H, Tamura Y, Go E, Kimura M, Kimura Y, Tsuchiya H, Yoshihara H, Hirokawa T, Endo T, Fon E, Trempe J-F, Saeki Y, Tanaka K, Matsuda N (2014) Ubiquitin is phosphorylated by PINK1 to activate parkin. Nature 510(7503):162–166
9. Shiba-Fukushima K, Arano T, Matsumoto G, Inoshita T, Yoshida S, Ishihama Y, Ryu KY, Nukina N, Hattori N, Imai Y (2014) Phosphorylation of mitochondrial polyubiquitin by PINK1 promotes parkin mitochondrial tethering. PLoS Genet 10(12):e1004861. doi:10.1371/journal.pgen.1004861
10. Ordureau A, Sarraf SA, Duda DM, Heo JM, Jedrychowski MP, Sviderskiy VO, Olszewski JL, Koerber JT, Xie T, Beausoleil SA, Wells JA, Gygi SP, Schulman BA, Harper JW (2014) Quantitative proteomics reveal a feedforward mechanism for mitochondrial PARKIN translocation and ubiquitin chain synthesis. Mol Cell. doi:10.1016/j.molcel.2014.09.007
11. Okatsu K, Koyano F, Kimura M, Kosako H, Saeki Y, Tanaka K, Matsuda N (2015) Phosphorylated ubiquitin chain is the genuine Parkin receptor. J Cell Biol 209(1):111–128. doi:10.1083/jcb.201410050
12. Greene JC, Whitworth AJ, Kuo I, Andrews LA, Feany MB, Pallanck LJ (2003) Mitochondrial pathology and apoptotic muscle degeneration in Drosophila parkin mutants. Proc Natl Acad Sci U S A 100(7):4078–4083. doi:10.1073/pnas.0737556100
13. Clark IE, Dodson MW, Jiang C, Cao JH, Huh JR, Seol JH, Yoo SJ, Hay BA, Guo M (2006) Drosophila pink1 is required for mitochondrial function and interacts genetically with parkin. Nature 441(7097):1162–1166. doi:10.1038/nature04779
14. Park J, Lee SB, Lee S, Kim Y, Song S, Kim S, Bae E, Kim J, Shong M, Kim JM, Chung J (2006) Mitochondrial dysfunction in Drosophila PINK1 mutants is complemented by parkin. Nature 441(7097):1157–1161. doi:10.1038/nature04788
15. Yang Y, Gehrke S, Imai Y, Huang Z, Ouyang Y, Wang JW, Yang L, Beal MF, Vogel H, Lu B (2006) Mitochondrial pathology and muscle and dopaminergic neuron degeneration caused by inactivation of Drosophila Pink1 is rescued by Parkin. Proc Natl Acad Sci U S A 103 (28):10793–10798. doi:10.1073/pnas.0602493103
16. Imai Y, Kanao T, Sawada T, Kobayashi Y, Moriwaki Y, Ishida Y, Takeda K, Ichijo H, Lu B, Takahashi R (2010) The loss of PGAM5 suppresses the mitochondrial degeneration caused by inactivation of PINK1 in Drosophila. PLoS Genet 6(12):e1001229. doi:10.1371/journal.pgen.1001229
17. Shiba-Fukushima K, Inoshita T, Hattori N, Imai Y (2014) PINK1-mediated phosphorylation of Parkin boosts Parkin activity in Drosophila. PLoS Genet 10(6):e1004391. doi:10.1371/journal.pgen.1004391

18. Tanaka A, Cleland MM, Xu S, Narendra DP, Suen DF, Karbowski M, Youle RJ (2010) Proteasome and p97 mediate mitophagy and degradation of mitofusins induced by Parkin. J Cell Biol 191(7):1367–1380. doi:10.1083/jcb.201007013
19. Ziviani E, Tao RN, Whitworth AJ (2010) Drosophila parkin requires PINK1 for mitochondrial translocation and ubiquitinates mitofusin. Proc Natl Acad Sci U S A 107(11):5018–5023. doi:10.1073/pnas.0913485107
20. Liu S, Sawada T, Lee S, Yu W, Silverio G, Alapatt P, Millan I, Shen A, Saxton W, Kanao T, Takahashi R, Hattori N, Imai Y, Lu B (2012) Parkinson's disease-associated kinase PINK1 regulates Miro protein level and axonal transport of mitochondria. PLoS Genet 8(3): e1002537. doi:10.1371/journal.pgen.1002537. PGENETICS-D-11-02331 [pii]
21. Wang X, Winter D, Ashrafi G, Schlehe J, Wong YL, Selkoe D, Rice S, Steen J, Lavoie MJ, Schwarz TL (2011) PINK1 and Parkin target Miro for phosphorylation and degradation to arrest mitochondrial motility. Cell 147 (4):893–906 . doi:10.1016/j.cell.2011.10.018S0092-8674(11)01224-4 [pii]
22. Yang Y, Ouyang Y, Yang L, Beal MF, McQuibban A, Vogel H, Lu B (2008) Pink1 regulates mitochondrial dynamics through interaction with the fission/fusion machinery. Proc Natl Acad Sci U S A 105(19):7070–7075. doi:10.1073/pnas.0711845105
23. Yang YF, Nishimura I, Imai Y, Takahashi R, Lu BW (2003) Parkin suppresses dopaminergic neuron-selective neurotoxicity induced by Pael-R in Drosophila. Neuron 37 (6):911–924. doi:10.1016/S0896-6273(03)00143-0

Methods in Molecular Biology (2018) 1759: 59–67
DOI 10.1007/7651_2017_10

Published online: 22 March 2017

Assessment of Mitophagy in iPS Cell-Derived Neurons

Kei-Ichi Ishikawa, Akihiro Yamaguchi, Hideyuki Okano, and Wado Akamatsu

Abstract

Aberrant mitochondrial function is associated with many neurological diseases. Mitophagy is a key mechanism for the elimination of damaged mitochondria and maintenance of mitochondrial homeostasis. Induced pluripotent stem (iPS) cell technologies developed over the last decade have allowed us to analyze functions of the human neuron. Here we describe an efficient induction method from human iPS cells to neurons, followed by an image-based mitophagy assay.

Keywords: Autophagy, iPS cells, Mitophagy, Neurons, Parkinson's disease

1 Introduction

The maintenance of intracellular organelles is important for healthy maintenance of neurons, in particular because of the limited turnover of adult primate neurons in certain brain regions. Neurons are supplied energy for their activity and survival primarily via oxidative phosphorylation, highlighting the importance of mitochondrial quality control for neuronal vitality. Increasing evidence indicates that mitochondrial abnormalities induce neurological diseases via factors such as excessive oxidative stress, abnormal calcium signaling, abnormal metabolism, and cell death [1, 2]. Mitophagy is a form of selective autophagy that eliminates damaged mitochondria. Thus, aberrant mitophagy function has been implicated in the pathogenesis of neurodegenerative diseases, such as Parkinson's disease, Alzheimer's disease, and Huntington's disease [3, 4].

Parkinson's disease is the second most common neurodegenerative disorder, characterized by degeneration of dopaminergic neurons in the substantia nigra. Mitochondrial dysfunction has been shown to play a major role in the pathogenesis of this disease. The cytosolic E3 ubiquitin ligase Parkin and PTEN-induced kinase 1 (PINK1) are key mitophagy proteins, and mutations in *parkin* and *PINK1* genes cause familial Parkinson's disease [5, 6]. Several other Parkinson's disease-related genes have also been shown to be associated with mitochondria or mitophagy [7], and abnormal mitochondria accumulate in the brains of Parkinson's disease patients [8].

In the recent decade, technological developments of induced pluripotent stem (iPS) cells and neuronal induction from iPS cells have allowed us to assess the functions of human neurons [9–11]. We and other groups have reported impaired mitophagy in iPS-derived neurons from patients with familial Parkinson's disease [11–13].

The present paper provides a protocol for highly efficient neuronal induction from on-feeder iPS cells, followed by an imaging-based mitophagy assay. Although western blotting is commonly used as a mitophagy assay in cultured cells, it is impossible to obtain highly purified neurons (~100%) from iPS cells by defined culture protocols including ours. Therefore, we believe imaging-based assays are more suitable for accurate analysis of mitophagy in iPS-derived neurons. This protocol could be applicable in drug-screening or pathophysiological analysis of Parkinson's disease, as well as other neurodegenerative diseases.

2 Materials

2.1 iPS Cell Culture and Neural Induction

1. Culture medium for human iPS cells (iPS medium): Dulbecco's Modified Eagle Medium/Nutrient Mixture F-12 (DMEM/F12) supplemented with 1% nonessential amino acid solution, 2 mM L-glutamine, 0.1 M 2-mercaptoethanol, 125 mL KnockOut Serum Replacement (KSR), 4 ng/mL fibroblast growth factor-2 (FGF-2), and 0.5% penicillin/streptomycin.
2. Culture medium for neurospheres (Neurosphere medium): KBM neural stem cell medium (Kohjin Bio) supplemented with B27, 20 ng/mL fibroblast growth factor-2 (FGF-2), 10 μM Y-27632, and 10 ng/mL human leukemia inhibitory factor (hLIF).
3. Culture medium for neurons (neuron medium): MHM/B27 medium supplemented with 10 ng/mL brain-derived neurotrophic factor (BDNF), 10 ng/mL glial cell-derived neurotrophic factor (GDNF), 1 mM dibutyryl-cAMP, and 200 μM ascorbic acid.
4. Dissociation solution for human ES/iPS cells (ReproCELL).
5. TrypLE Select (Life Technologies).
6. 0.1% gelatin-coated 10-cm dish: 10-cm culture plate previously coated with 0.1% gelatin at 37°C overnight.
7. Trypsin inhibitor (Sigma).
8. Poly-L-ornithine (150 μg/mL stock); use at 15 μg/mL.

9. Fibronectin (0.5 mg/mL stock); use at 10 μg/mL.
10. Culture plates (10 cm, 48-well).
11. T75 flasks.
12. Glass coverslips (10 mm in diameter).
13. Slide glasses.
14. 70-μm cell strainer.
15. Phosphate-buffered saline (PBS).

2.2 Image-Based Analysis of Mitophagy

1. Carbonyl cyanide m-chlorophenyl hydrazone (CCCP); use at 30 μM.
2. Bafilomycin A1 (BafA1); use at 5 μM.
3. Dimethyl sulfoxide (DMSO).
4. Paraformaldehyde (PFA).
5. Blocking buffer: 5% (v/v) fetal bovine serum (FBS), 0.5% (v/v) Tween 20 in PBS.
6. Primary antibodies: anti-βIII-tubulin (Sigma, neuronal marker), anti-complex III core I (Thermo Fisher Scientific, mitochondrial inner membrane marker).
7. Secondary antibodies: Alexa Fluor 488- and Alexa Fluor 555-conjugated secondary antibodies (Thermo Fisher Scientific).
8. Mounting medium.
9. Nail polish.

3 Methods

3.1 Induction of Neurons from Human iPS Cells

The overview of this protocol is shown in Fig. 1.

1. Culture human iPS cells on feeder cells in a 10-cm dish. Aspirate the medium and wash once in 10 mL PBS (*see* **Note 1**).
2. Treat with 1 mL dissociation solution and aspirate. Incubate the dish at 37°C for 3 min.
3. Add 10 mL PBS and aspirate PBS with floating feeder cells (*see* **Note 2**).
4. Add 10 mL iPS medium and detach the iPS colonies with a cell scraper. Collect the colonies in 15-mL tube.
5. Centrifuge for 5 min at 200 × g and aspirate the supernatant.
6. Mildly resuspend the colonies with 10 mL iPS cell medium containing 10 μM Y-27632, and transfer to the gelatin-coated 10-cm dish.
7. Incubate for 2–4 h, at 37°C, 3% CO_2.

iPS cells cultured on feeder cells

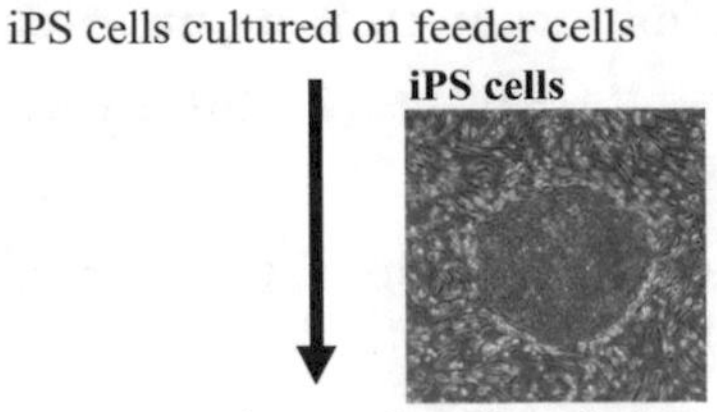

Strip feeder cells away by treating dissociation solution

Incubate on a gelatin coated dish to remove feeder cells

Dissociate iPS cell colonies into single cells by treating T rypLE Select and pipetting

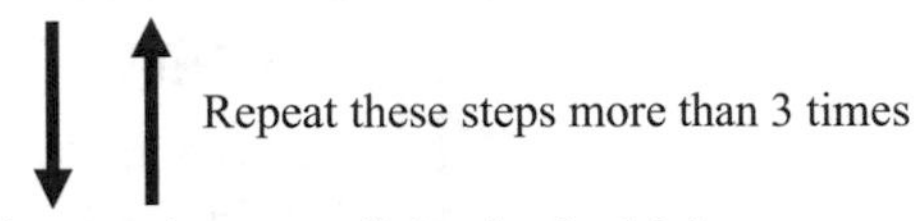

Incubate in Neurosphere medium at adequate cell density for 14 days

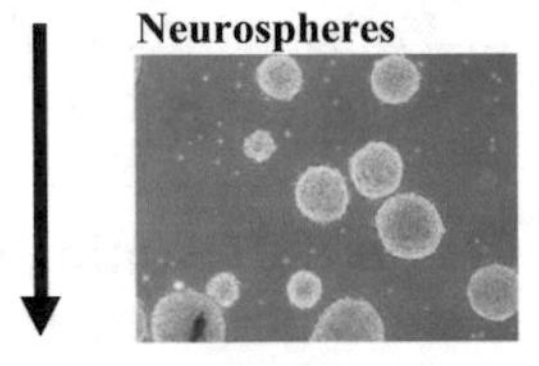

Incubate in Neuron medium on the poly-ornitine/fibronectin coated dishes for 14 days

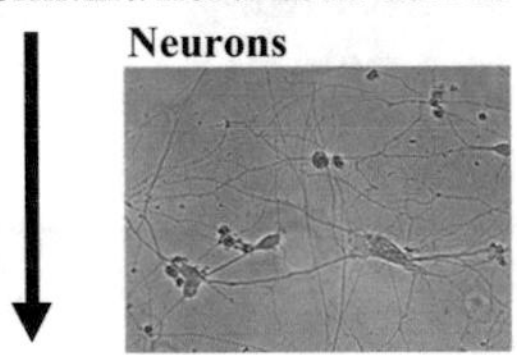

Mitophagy assay
(Treat with CCCP and BafA1 followed by immunofluorescence)

Fig. 1 Overview flowchart of neural induction from on-feeder iPS cells

8. Collect iPS medium with the floating iPS cell colonies in a 50-mL tube (*see* **Note 3**).
9. Centrifuge for 5 min at 200 × g and aspirate the supernatant.
10. Add 1 mL TrypLE Select and incubate in a 37°C water bath for 5 min.
11. Triturate cells using a P1000 pipette about 10 times to suspend (*see* **Note 4**).
12. Add 2 mL trypsin inhibitor and filtrate through a 70-μm cell strainer.
13. Wash the cell strainer with 7 mL Neurosphere medium.
14. Centrifuge for 5 min at 200 × g and aspirate the supernatant.

15. Resuspend cells with 2 mL Neurosphere medium and quantify the cells.
16. Add 4×10^5 cells in a T75 flask containing 40 mL Neurosphere medium.
17. Incubate for 14 days at 37°C, 5% CO_2, 4% O_2 (*see* **Note 5**).
18. After 14 days, the suspended cells that form spheres of colonies are the primary neurospheres.
19. Collect the neurospheres in a 50-mL tube.
20. Centrifuge for 5 min at $200 \times g$ and aspirate the supernatant.
21. To establish secondary neurospheres, follow the same procedure (steps #19–20 and #10–17) except the cell density should be 2×10^6/40 mL in step #16.
22. To establish tertiary neurospheres, follow the same procedure as steps #19–21.
23. Place cover glasses in more than four wells of a 48-well dish, and coat the dishes with 15 μg/mL poly-L-ornithine then 10 μg/mL fibronectin overnight.
24. For neural induction, follow steps #19–20 and #10–15.
25. Discard the coating solution in the 48-well dish. Add 500 μL neuron medium, and plate the cells at a density of 9×10^4 cells/well.
26. Change the neuron medium every 3 days for 14 days (*see* **Note 6**).

3.2 Mitophagy Assessment

1. Treat the four wells of cultured neurons with DMSO, 30 μM CCCP, DMSO + 5 μM BafA1, and 30 μM CCCP + 5 μM BafA1 for 48 h at 37°C.
2. For fixation, add the same amount of 8% PFA, and incubate for 10 min at room temperature (RT).
3. Aspirate the liquid, and incubate with 4% PFA for 20 min at room temperature (RT).
4. Wash the cells three times in PBS.
5. Boil pure water in a plastic case in a microwave oven, and place a dish with fixed samples in the boiled water for 5 min (*see* **Note 7**).
6. Wash cells once in PBS.
7. Incubate the cells with blocking buffer for 30 min at RT for permeabilization and blocking.
8. Aspirate the blocking buffer and the incubate cells with anti-complex III core I (1:500) and anti-βIII-tubulin (1:1,000) antibodies in blocking buffer overnight at 4°C.
9. Wash the cells three times in PBS for 5 min each.

10. Incubate with secondary antibodies (1:500) and Hoechst (1:10,000) in blocking buffer for 1 h at RT in the dark.
11. Wash the cells three times in PBS for 5 min each.
12. Remove the coverslips from the wells. Place mounting medium onto the coverslips, and place them facedown onto the glass slides.
13. Seal the coverslip with nail polish.
14. Mitochondrial images in neurons are obtained by fluorescence microscopy or confocal microscopy (Fig. 2a).
15. Measure the mitochondrial area (complex III core I expression) in the neuronal cell soma (βIII-tubulin expression) using imaging analysis software, such as Image J (*see* **Notes 8** and **9**).
16. The ratio of average mitochondrial area in a cell treated with CCCP to DMSO and the ratio of average area of CCCP + BafA1 to DMSO + BafA1 can be used to compare the number of mitochondria eliminated by mitophagy (*see* **Note 10**) (Fig. 2b).

4 Notes

1. Protocols for generation and maintenance of human iPS cells are described by Ohnuki et al. [14].
2. Only feeder cells detach from the dish, and iPS cell colonies remain attached to the dish.
3. Almost all feeder cells attach to the dish in this step. Floating iPS cell colonies should be collected.
4. Pipette carefully until the colonies are dissociated into single cells. This step is important for obtaining sufficient numbers of cells.
5. Colony formation is visible by day 3. Adjust the incubation period from 10 to 20 days depending on colony growth.
6. We usually start the mitophagy assay at day 14, when there is sufficient neuronal growth. Depending on the purpose of the experiments, the incubation period can be extended to be able to perform the assay with more mature neurons.
7. This antigen retrieval step is necessary for binding of the anti-compIII-coreI antibody. However, you may skip this step when you use other antibodies specific for mitochondrial proteins. The mitochondrial proteins become degraded by mitophagy and the ubiquitin-proteasome system. Therefore, we use an antibody specific for the mitochondrial inner membrane protein, complex III core I [15].

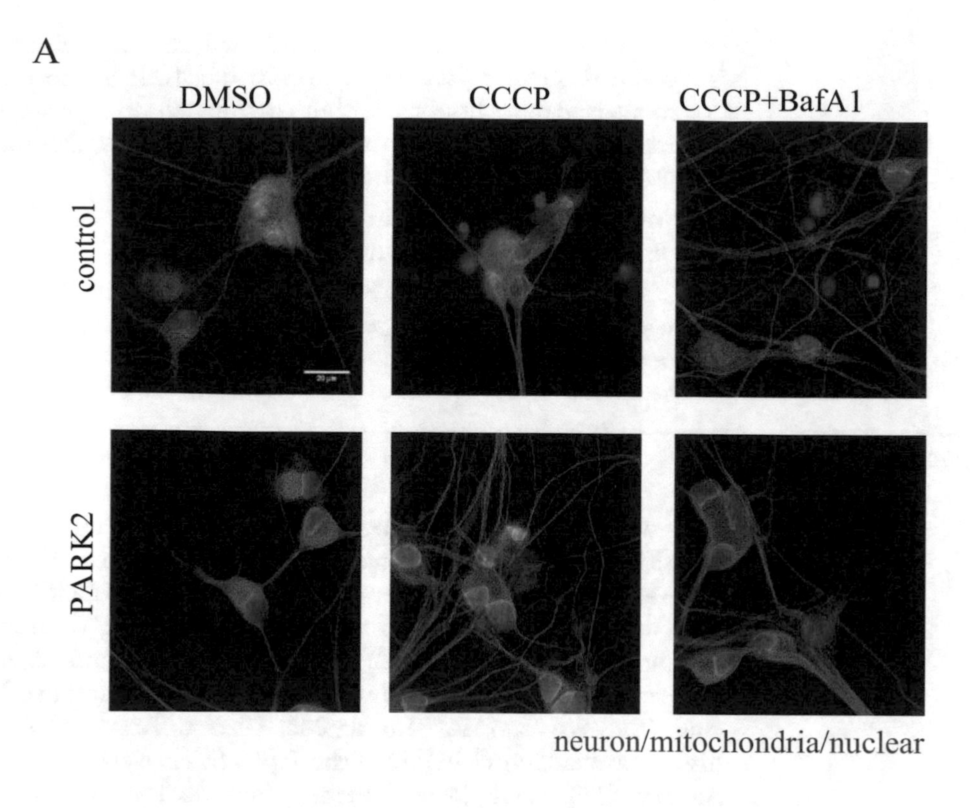

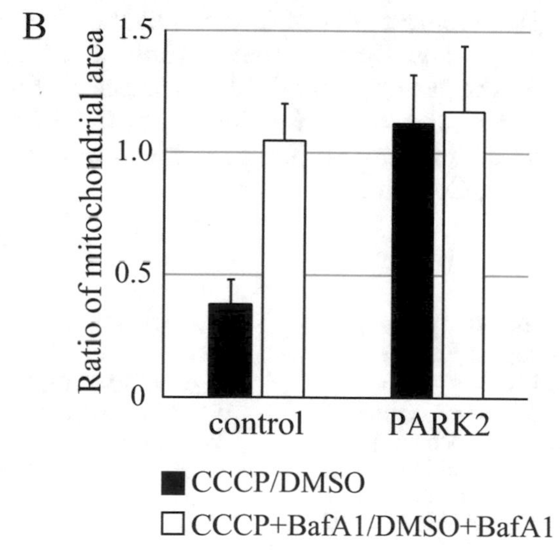

Fig. 2 Analysis of mitophagy neurons derived from iPS cells. Control- and PARK2-neurons were treated with DMSO, 30 μM CCCP, and/or 5 μM BafA1 for 48 h. Neurons were immunostained with complex III core I and βIII-tubulin antibodies (**a**) and the areas of mitochondria as quantified (**b**). Mitochondria are not eliminated in CCCP-treated PARK2 neurons owing to a mutation in the *parkin* gene. Bar, 20 μm. Error bars represent SEM

8. To exclude non-neuronal cells, analyze mitochondria in comparison with βIII-tubulin-positive cells. It is also possible to analyze mitochondria within specific neuronal populations, such as dopaminergic neurons or motor neurons, through the use of a specific neuronal marker.
9. We recommend analyzing >30 neurons per condition to obtain statistical significance.
10. The difference in ratios between conditions with BafA1 and without BafA1 represents the amount of mitochondria eliminated by mitophagy.

Acknowledgments

This work was supported by JSPS KAKENHI Grant Number JP16K19524 to K.I.; the Project for the Realization of Regenerative Medicine and Support for Core Institutes for iPS Cell Research from the Ministry of Education, Culture, Sports, Science and Technology of Japan (MEXT) to H.O.; Research Center Network for Realization Research Centers/Projects of Regenerative Medicine (the Program for Intractable Disease Research utilizing disease-specific iPS Cells) from the Japan Science and Technology Agency (JST) and Japan Agency for Medical Research and Development (AMED) to H.O.; the New Energy and Industrial Technology Development Organization (NEDO) to H.O. and W. A.; the Japan Society for the Promotion of Science (JSPS) to W.A.; and a Grant-in-Aid for the Global COE Program from MEXT to Keio University. H.O. is a scientific consultant for SanBio, Co. Ltd., Eisai, Co., Ltd.

References

1. Arun S, Liu L, Donmez G (2016) Mitochondrial biology and neurological diseases. Curr Neuropharmacol 14(2):143–154
2. Celsi F, Pizzo P, Brini M, Leo S, Fotino C, Pinton P, Rizzuto R (2009) Mitochondria, calcium and cell death: a deadly triad in neurodegeneration. Biochim Biophys Acta 1787 (5):335–344. doi:10.1016/j.bbabio.2009.02.021
3. Palikaras K, Tavernarakis N (2012) Mitophagy in neurodegeneration and aging. Front Genet 3:297. doi:10.3389/fgene.2012.00297
4. Pellegrino MW, Haynes CM (2015) Mitophagy and the mitochondrial unfolded protein response in neurodegeneration and bacterial infection. BMC Biol 13:22. doi:10.1186/s12915-015-0129-1
5. Hattori N, Saiki S, Imai Y (2014) Regulation by mitophagy. Int J Biochem Cell Biol 53:147–150. doi:10.1016/j.biocel.2014.05.012
6. Pickrell AM, Youle RJ (2015) The roles of PINK1, parkin, and mitochondrial fidelity in Parkinson's disease. Neuron 85(2):257–273. doi:10.1016/j.neuron.2014.12.007
7. Beilina A, Cookson MR (2015) Genes associated with Parkinson's disease: regulation of autophagy and beyond. J Neurochem. doi:10.1111/jnc.13266
8. Celardo I, Martins LM, Gandhi S (2014) Unravelling mitochondrial pathways to Parkinson's disease. Br J Pharmacol 171 (8):1943–1957. doi:10.1111/bph.12433
9. Okano H, Yamanaka S (2014) iPS cell technologies: significance and applications to CNS regeneration and disease. Mol Brain 7:22. doi:10.1186/1756-6606-7-22

10. Imaizumi K, Sone T, Ibata K, Fujimori K, Yuzaki M, Akamatsu W, Okano H (2015) Controlling the regional identity of hPSC-derived neurons to uncover neuronal subtype specificity of neurological disease phenotypes. Stem Cell Reports 5(6):1010–1022. doi:10.1016/j.stemcr.2015.10.005
11. Matsumoto T, Fujimori K, Andoh-Noda T, Ando T, Kuzumaki N, Toyoshima M, Tada H, Imaizumi K, Ishikawa M, Yamaguchi R, Isoda M, Zhou Z, Sato S, Kobayashi T, Ohtaka M, Nishimura K, Kurosawa H, Yoshikawa T, Takahashi T, Nakanishi M, Ohyama M, Hattori N, Akamatsu W, Okano H (2016) Functional neurons generated from T cell-derived induced pluripotent stem cells for neurological disease modeling. Stem Cell Reports 6(3):422–435. doi:10.1016/j.stemcr.2016.01.010
12. Seibler P, Graziotto J, Jeong H, Simunovic F, Klein C, Krainc D (2011) Mitochondrial Parkin recruitment is impaired in neurons derived from mutant PINK1 induced pluripotent stem cells. J Neurosci Off J Soc Neurosci 31 (16):5970–5976. doi:10.1523/JNEUROSCI.4441-10.2011
13. Imaizumi Y, Okada Y, Akamatsu W, Koike M, Kuzumaki N, Hayakawa H, Nihira T, Kobayashi T, Ohyama M, Sato S, Takanashi M, Funayama M, Hirayama A, Soga T, Hishiki T, Suematsu M, Yagi T, Ito D, Kosakai A, Hayashi K, Shouji M, Nakanishi A, Suzuki N, Mizuno Y, Mizushima N, Amagai M, Uchiyama Y, Mochizuki H, Hattori N, Okano H (2012) Mitochondrial dysfunction associated with increased oxidative stress and alpha-synuclein accumulation in PARK2 iPSC-derived neurons and postmortem brain tissue. Mol Brain 5:35. doi:10.1186/1756-6606-5-35
14. Ohnuki M, Takahashi K, Yamanaka S (2009) Generation and characterization of human induced pluripotent stem cells. Curr Protoc Stem Cell Biol Chapter 4:Unit 4A 2. doi:10.1002/9780470151808.sc04a02s9
15. Yoshii SR, Kishi C, Ishihara N, Mizushima N (2011) Parkin mediates proteasome-dependent protein degradation and rupture of the outer mitochondrial membrane. J Biol Chem 286(22):19630–19640. doi:10.1074/jbc.M110.209338

Part II

Mitophagy Associated with Nix

Methods in Molecular Biology (2018) 1759: 71–83
DOI 10.1007/7651_2017_11

Published online: 24 March 2017

Investigation of Yeast Mitophagy with Fluorescence Microscopy and Western Blotting

Sachiyo Nagumo and Koji Okamoto

Abstract

Selective clearance of superfluous or dysfunctional mitochondria is a fundamental process that depends on the autophagic membrane trafficking pathways found in many cell types. This catabolic event, called mitophagy, is conserved from yeast to humans and serves to control mitochondrial quality and quantity. In budding yeast, degradation of mitochondria occurs under various physiological conditions, such as respiration at stationary phase, or starvation in a prolonged period. During these events, the transmembrane protein Atg32 localizes to the mitochondrial surface and plays a specific and essential role in yeast mitophagy. In this chapter, we describe methods to observe transport of mitochondria to the vacuole, a lytic compartment in yeast, using fluorescence microscopy, and semi-quantify the progression of Atg32-mediated mitophagy by Western blotting.

Keywords: Atg32, Fluorescence microscopy, Mitophagy, Western blotting, Yeast

1 Introduction

Mitochondria are highly dynamic organelles that frequently change their shape and volume in response to environmental cues and cellular metabolic states [1]. Mitochondrial form is determined predominantly by their opposing *fission* and *fusion* events, while mitochondrial amount is established primarily by a balance between their *biogenesis* and *degradation* [2, 3]. These four processes also contribute to the maintenance of mitochondrial activities, as the power plants of the cell are constantly challenged with their own by-products, reactive oxygen species generated through the electron transport chain [4, 5]. In particular, recent studies reveal that autophagy-dependent degradation selective for mitochondria, termed mitophagy, acts as a mitochondrial quantity and quality control mechanism and that mitophagic failure is linked to manifold disorders including respiratory deficiency, aberrant cell differentiation, and neurodegeneration, underscoring the biological relevance of this catabolic pathway [6].

Mitophagy has so far been found among various eukaryotes and classified into ubiquitin-dependent and ubiquitin-independent processes [7]. In the former case, best known as the parkin/PINK1-mediated mitophagy in mammalian cells [8, 9], ubiquitin links covalently to multiple proteins on the mitochondrial surface and serves as a degradation tag that binds adaptor proteins recruiting the autophagy machinery to mitochondria. In the latter case, a specific outer membrane-anchored protein functions as a receptor that interacts with the Atg8 family member and other proteins required for sequestration of mitochondria [10]. Notably, mitophagy receptors, such as Atg32 in yeast and NIX, BNIP3, FUNDC1, and Bcl2-L-13 in mammals, share little sequence similarity but promote mitochondrial clearance via a common mechanism [11, 12]. Supporting this idea, a recent study has demonstrated that Bcl2-L-13 expressed ectopically in yeast cells lacking Atg32 can drive autophagy-dependent degradation of mitochondria [12]. Thus, studies using yeast provide valuable insights into the basic principles underlying receptor-mediated mitophagy in mammals.

The budding yeast *Saccharomyces cerevisiae* undergoes drastic mitophagy in a manner dependent on Atg32 when the cells are cultured in medium containing a non-fermentable carbon source (e.g., glycerol, ethanol, or lactate) for a prolonged period [13, 14]. Under this condition, mitochondrial respiration is essential for cell viability, and oxidative phosphorylation becomes highly active during cell growth. Upon entry into stationary phase, a substantial fraction of mitochondria is selectively sequestered into double membrane-bound structures called autophagosomes and transported to the vacuole (lysosome). In addition, mitophagy occurs when cells are shifted from nutrient-rich to nitrogen-deprived conditions in medium containing a fermentable carbon source such as glucose [15]. This catabolic process is distinct from bulk autophagy, as the former depends on Atg32 and takes place significantly later than the latter. Notably, starvation-induced mitophagy does not require mitochondrial energy metabolism, which thus makes it possible to assess degradation of mitochondria even in respiratory-deficient mutants.

This chapter provides two conventional protocols to investigate receptor-mediated mitophagy in yeast under prolonged respiration or starvation. First, using fluorescence microscopy, transport of mitochondria to the vacuole can be visualized in living cells. Second, the progression of mitochondrial degradation can be measured semiquantitatively by Western blotting. Expression of a mitophagy marker allows one to use these methods and obtain the data from the same cell population. In combination with powerful yeast genetics and high-throughput techniques, these approaches may be applicable to screens for chemical compounds that specifically facilitate or inhibit degradation of mitochondria.

2 Materials

2.1 Generation of DNA Cassette for Yeast Transformation

1. Plasmid: pBSII-TEF^{p}-mito-DHFR-mCherry-CgHIS3 (*see* **Note 1**). This plasmid can be requested and sent from us.
2. Primers for PCR amplification of the TEF^{p}-mito-DHFR-mCherry-CgHIS3 cassette that can be integrated into the *his3* locus: forward primer, 5′-AATGTGATTTCTTCGAAGAATA-TACTAAAAAATGAGCAGGCAAGATAAACGAAGGCAAA-GCGACGGTATCGATAAGCTTG-3′; reverse primer, 5′-GGTATACATATATACACATGTATATATATCGTATGCTG-CAGCTTTAAATAATCGGTGTCACAAGCGCGCAATTAA CCCTC-3′. These long primers should be purified by an oligonucleotide purification cartridge. The concentration of these primers is 0.1 mM.

2.2 Construction of Yeast Strains for Mitophagy Assays

1. Yeast strain: *S. cerevisiae* BY4741, *MATa his3Δ1 lue2Δ0 met15Δ ura3Δ* (*see* **Note 2**).
2. Rich liquid medium supplemented with dextrose (YPD): 1% (w/v) yeast extract, 2% (w/v) peptone, and 2% (w/v) dextrose. YP medium (yeast extract, peptone) and dextrose are separately dissolved in water, and mix them after autoclaving. Store at room temperature.
3. Synthetic solid medium supplemented with dextrose and histidine dropout (SD-His): 0.17% (w/v) yeast nitrogen base without amino acids and ammonium sulfate, 0.5% (w/v) ammonium sulfate, 2% (w/v) dextrose, amino acid/nucleoside solutions without histidine, and 2% (w/v) agar. S medium (yeast nitrogen base, ammonium sulfate, agar) and dextrose are separately dissolved in water, and mix together after autoclaving. Supplement them with filter-sterilized amino acid/nucleoside solutions without histidine. Store at 4°C.
4. Sterile water.
5. 1 M lithium acetate (LiAc). Filter sterilize and store at room temperature.
6. 0.1 M LiAc. Filter sterilize and store at room temperature.
7. 60% (w/v) polyethylene glycol (PEG) (average molecular weight: 3,350). Filter sterilize and store at room temperature.
8. 10 mg/mL single-stranded salmon sperm DNA (ssDNA). Boil for 5 min and store at −20°C.
9. TEF^{p}-mito-DHFR-mCherry-CgHIS3 DNA cassette (*see* Section 3.1). Store at −20°C.

2.3 Fluorescence Microscopy for Live Cell Imaging of Mitophagy

1. BY4741 *his3Δ1::TEF^P-mito-DHFR-mCherry::CgHIS3* (positive control strain) [15].
2. BY4741 *his3Δ1::TEF^P-mito-DHFR-mCherry::CgHIS3 atg32Δ* (negative control strain) [15].
3. Synthetic liquid medium supplemented with dextrose and casamino acids (SDCA): 0.17% (w/v) yeast nitrogen base without amino acids and ammonium sulfate, 0.5% (w/v) ammonium sulfate, 0.5% (w/v) casamino acids, 2% (w/v) dextrose, 20 μg/mL adenine sulfate, 20 μg/mL L-tryptophan, and 20 μg/mL uracil. SCA medium (yeast nitrogen base, ammonium sulfate, casamino acids) and dextrose are separately dissolved in water, and mix together after autoclaving. Supplement them with filter-sterilized amino acid/nucleoside solutions and store at room temperature.
4. Synthetic liquid medium supplemented with glycerol and casamino acids (SDGlyCA): 0.17% (w/v) yeast nitrogen base without amino acids and ammonium sulfate, 0.5% (w/v) ammonium sulfate, 0.5% (w/v) casamino acids, 0.1% (w/v) dextrose, 3% (v/v) glycerol, 20 μg/mL adenine sulfate, 20 μg/mL L-tryptophan, and 20 μg/mL uracil. SCA medium (yeast nitrogen base, ammonium sulfate, casamino acids), dextrose, and glycerol are separately dissolved in water, and mix together after autoclaving. Supplement them with filter-sterilized amino acid/nucleoside solutions and store at room temperature.
5. YPD: *see* Section 2.2, **item 2**.
6. Nitrogen-deprived synthetic liquid medium supplemented with dextrose (SD-N): 0.17% (w/v) yeast nitrogen base without amino acids and ammonium sulfate and 2% (w/v) dextrose. Yeast nitrogen base and dextrose are separately dissolved in water, and mix together after autoclaving. Store at room temperature.

2.4 Western Blotting for Semiquantification of Mitophagy

1. BY4741 *his3Δ1::TEF^P-mito-DHFR-mCherry::CgHIS3* (*see* Section 2.3, **item 1**).
2. BY4741 *his3Δ1::TEF^P-mito-DHFR-mCherry::CgHIS3 atg32Δ* (*see* Section 2.3, **item 2**).
3. SDCA: *see* Section 2.3, **item 3**.
4. SDGlyCA: *see* Section 2.3, **item 4**.
5. YPD: *see* Section 2.2, **item 2**.
6. SD-N: *see* Section 2.3, **item 6**.
7. 0.1 *M* NaOH. Store at room temperature.

8. Sample buffer: 60 mM Tris-HCl, pH 6.8, 5% (v/v) glycerol, 2% (w/v) SDS, 20 mM DTT, and 0.005% (w/v) bromophenol blue. Store at 4°C.
9. Polyacrylamide gel containing 10% (w/v) acrylamide/bis-acrylamide (ratio 37.5:1)
10. Polyvinylidene fluoride (PVDF) membrane (pore size: 0.45 μm): Immobilon-P Membrane (IPVH00010, EMD Millipore). Activate in methanol before use.
11. Blocking buffer: TBS-T (*see* Section 2.4, **item 14**), 5% (w/v) skim milk.
12. Incubation buffer A for primary antibody: TBS-T, 0.1% (w/v) skim milk.
13. Incubation buffer B for secondary antibody: TBS-T, 1% (w/v) skim milk.
14. Washing buffer (Tris-buffered saline with Tween 20, TBS-T): 25 mM Tris-HCl, pH 7.4, 138 mM NaCl, 2.7 mM KCl, and 0.1% Tween-20.
15. Primary antibody: mouse monoclonal anti-mCherry antibody ([1C51] ab125096, Abcam), mouse monoclonal anti-Por1 antibody ([16G9E6BC4] ab110326, Abcam), and mouse monoclonal anti-Pgk1 antibody ([22C5D8] ab113687, Abcam). Store at −20°C.
16. Secondary antibody: horseradish peroxidase (HRP)-conjugated rabbit polyclonal anti-mouse IgG (H+L) (315-035-003, Jackson ImmunoResearch). Store at −20°C.
17. Enhanced chemiluminescence (ECL) detection regents: Western Lightning Plus-ECL (NEL103001EA, PerkinElmer). Store at 4°C.

3 Methods

3.1 Generation of DNA Cassette for Yeast Transformation

The TEF^{P}-mito-DHFR-mCherry-CgHIS3 DNA cassette (3.2 kb) can be generated by PCR amplification (*see* a standard reaction below) with the template plasmid and primers (*see* Section 2.1, **items 1 and 2**), purified using a spin column, and eluted into sterile water. The resulting DNA solution is typically at a concentration of 0.2–0.6 mg/mL.

1. Prepare a PCR reaction mixture (PCR buffer for KOD-Plus-Neo (KOD-401, TOYOBO), 0.2 mM dNTPs, 1.5 mM $MgSO_4$, 0.3 μM forward/reverse primers, 200 ng/mL template plasmid, 20 U/mL KOD-Plus-Neo). A typical reaction volume is 200 μL (four aliquots of 50 μL in 0.2-mL PCR tubes).

2. Perform a reaction using a thermal cycler with the following program: 1 cycle of (1) 94°C, 2 min; 30 cycles of (2) 98°C, 10 s, (3) 55°C, 30 s, and (4) 68°C, 3 min; 1 cycle of (5) 68°C, 10 min; followed by (6) 4°C, ever.
3. Evaluate the amplified DNA by agarose gel electrophoresis.
4. Purify the sample using a spin column and elute into sterile water.

3.2 Construction of Yeast Strains for Mitophagy Assays

The PCR-amplified TEF^{P}-mito-DHFR-mCherry-CgHIS3 DNA cassette, which contains terminal 60 nucleotides derived from the upstream and downstream of the *his3Δ1* locus, can be introduced into yeast cells by a standard LiAc/ssDNA/PEG method. The introduced cassette is then inserted into the *his3Δ1* locus by homologous recombination. The transformants can be selected with histidine prototrophy.

1. Grow yeast cells to mid-log phase (OD_{600} = 0.8–1.2) in 3 mL YPD at 30°C (*see* **Note 3**).
2. Transfer cells (0.5 OD_{600} units) to a 1.5-mL microcentrifuge tube, harvest them by centrifugation (16,000 *g*, 15 s), and discard the supernatant using a micropipette.
3. Wash the cells once with 500 μL sterile water, harvest them by centrifugation (16,000 *g*, 15 s), and discard the supernatant using a micropipette.
4. Resuspend the cells in 400 μL 0.1 M LiAc and incubate at 30°C for 10 min.
5. Harvest the cells by centrifugation (16,000 *g*, 15 s) and discard the supernatant completely using a micropipette.
6. Add 68.5 μL 60% (w/v) PEG, 10.2 μL 1 M LiAc, 1.4 μL 10 mg/mL ssDNA, 10 μL 0.4 mg/mL DNA cassette, and 10 μL sterile water to the cell pellet in the order listed and vortex vigorously.
7. Incubate the cells at 30°C for 30 min, followed by heat shock at 42°C for 20 min.
8. Harvest the cells by centrifugation (16,000 *g*, 15 s) and discard the supernatant using a micropipette.
9. Add 50 μL sterile water to the cell pellet and vortex vigorously.
10. Spread the cell suspension on a solid SD-His agar plate and incubate at 30°C for 3–4 days.
11. Restreak the colonies on flesh SD-His plates and incubate at 30°C for 1–2 days.
12. Grow the transformed cells to mid-log phase in 3 mL SDCA and confirm expression of mito-DHFR-mCherry using fluorescence microscopy and Western blotting (*see* **Note 4**).

3.3 Fluorescence Microscopy for Live Cell Imaging of Mitophagy

Transport of mitochondria to the vacuole in yeast cells expressing the mitophagy marker mito-DHFR-mCherry can be examined under a fluorescence microscope (*see* Fig. 1A, B) [15]. When cells under respiratory conditions are grown to late-log phase (Gly 24 h), mito-DHFR-mCherry indicates mitochondrial short tubules and fragments. Strikingly, cells at stationary phase (Gly 72 h) exhibit Atg32-dependent formation of spherical red fluorescence

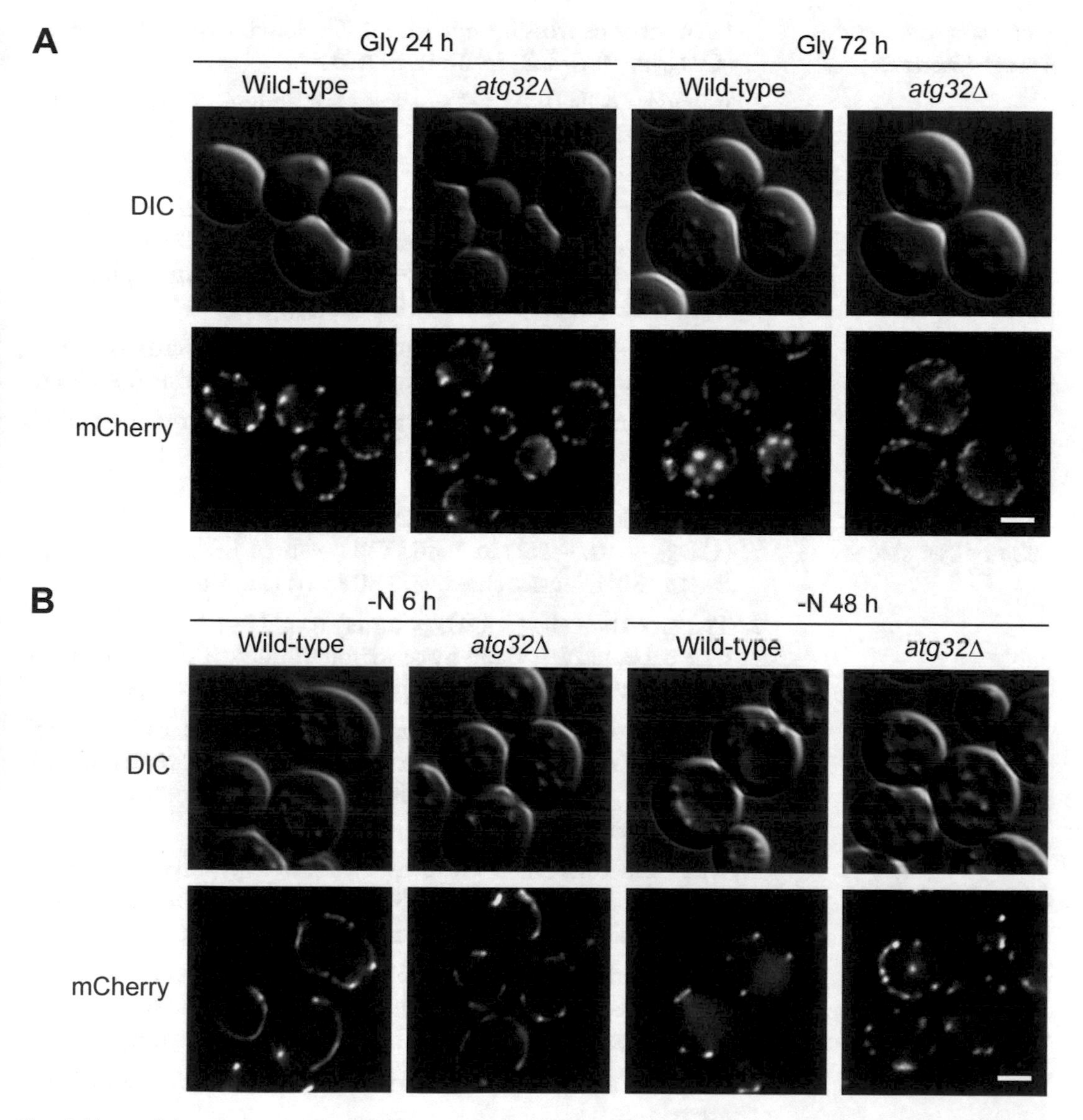

Fig. 1 Live cell imaging for mitophagy in yeast. (**A**) Wild-type and *atg32*-null cells expressing mito-DHFR-mCherry were grown under respiratory conditions (Gly), collected at the indicated time points, and observed using fluorescence microscopy. *DIC* differential interference contrast. Scale bar, 2 μm. (**B**) Wild-type and *atg32*-null cells expressing mito-DHFR-mCherry were pregrown under nutrient-rich conditions, shifted to nitrogen starvation (−N), collected at the indicated time points, and observed using fluorescence microscopy. *DIC* differential interference contrast. Scale bar, 2 μm

patterns that are typical to vacuolar localizations (*see* **Note 5**). Similarly, mitophagy is induced by prolonged starvation in a manner dependent on Atg32. When cells undergo drastic bulk autophagy (−N 6 h), mitochondria keep forming elongated tubules. Subsequently, cells under long-term nitrogen deprivation display red fluorescence patterns indicating not only mitochondrial short tubules and fragments but also vacuolar spheres (−N 48 h).

3.3.1 Mitophagy Under Prolonged Respiration

1. Grow cells expressing mito-DHFR-mCherry to mid-log phase (OD_{600} = 0.8–1.2) in 3 mL SDCA.
2. Inoculate cells (0.07 OD_{600} units) directly to 7 mL SDGlyCA in a 50-mL Erlenmeyer flask, and incubate them at 30°C under shaking (180 rpm).
3. Transfer the cells (0.5 OD_{600} units) to a 1.5-mL microcentrifuge tube, harvest them by centrifugation (16,000 *g*, 15 s), and discard a portion of the supernatant using a micropipette to leave 5 μL cell suspension (100 OD_{600}/mL).
4. Pipet 1.6–2.0 μL of the cell suspension onto a microscope slide, and cover with an 18-mm^2 square cover slip without sealing.
5. View the cells using a fluorescence microscope equipped with a filter set for mCherry.

3.3.2 Mitophagy Under Prolonged Starvation

1. Grow cells expressing mito-DHFR-mCherry to mid-log phase (OD_{600} = 0.8–1.2) in 7 mL YPD using a 50-mL Erlenmeyer flask at 30°C under shaking (180 rpm) (*see* **Note 6**).
2. Transfer the cells (7 OD_{600} units) to a 15-mL conical centrifuge tube, harvest them by centrifugation (1,600 *g*, 3 min), and discard the supernatant completely using a micropipette.
3. Resuspend the cells in 1 mL SD-N, transfer to a 1.5-mL microcentrifuge tube, harvest them by centrifugation (16,000 *g*, 15 s), and discard the supernatant using a micropipette.
4. Wash the cells once more in 1 mL SD-N, harvest them by centrifugation (16,000 *g*, 15 s), and discard the supernatant using a micropipette (*see* **Note 7**).
5. Resuspend the cells in 1 mL SD-N, transfer the cell suspension to 6 mL SD-N (1 OD_{600}/mL) in a 50-mL Erlenmeyer flask, and incubate them at 30°C under shaking (180 rpm).
6. Prepare a sample for observation as described in Section 3.3.1, **steps 3–5**.

3.4 Western Blotting for Semiquantification of Mitophagy

In addition to its utility for fluorescence imaging, the mito-DHFR-mCherry marker serves to monitor degradation of mitochondria by Western blotting (*see* Fig. 2A, B) [15]. When cells undergo mitophagy, this mitochondria-localized marker is processed to generate

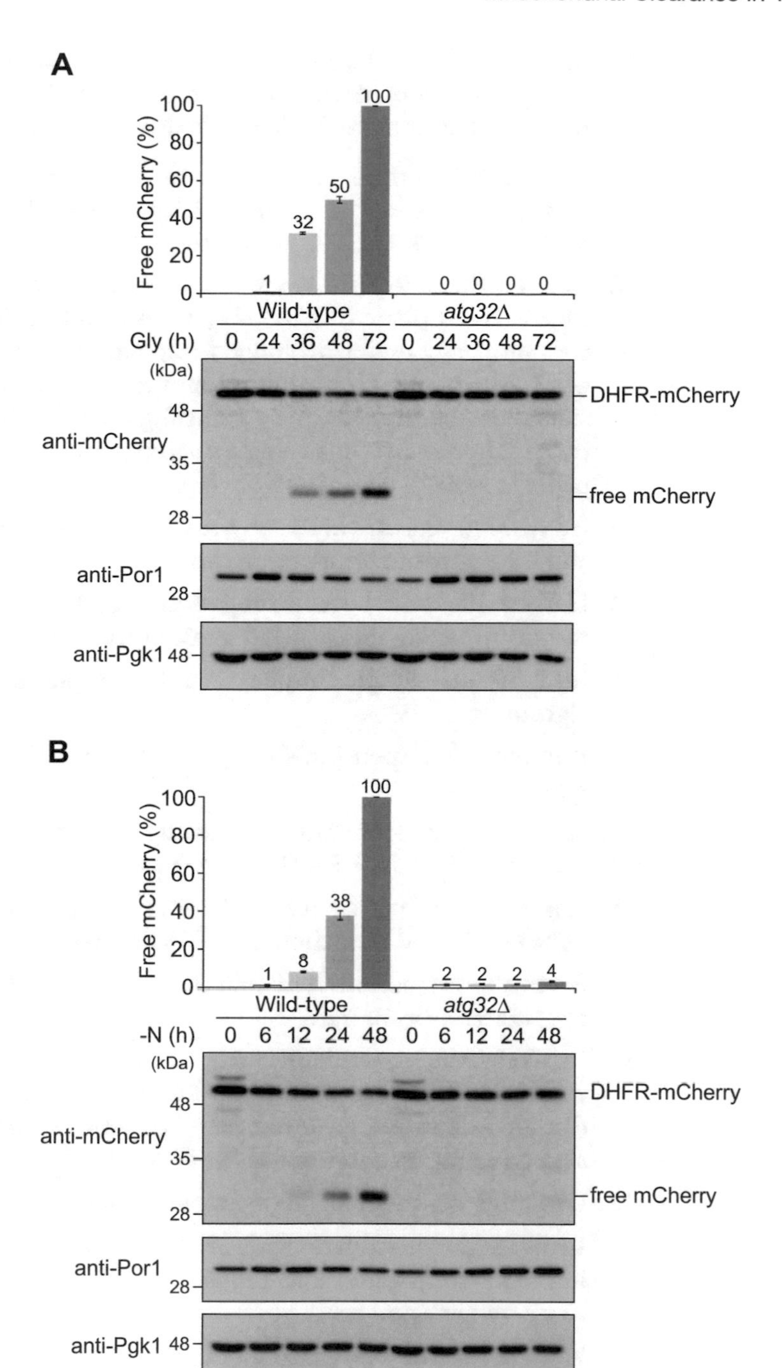

Fig. 2 Western blot analysis for mitophagy in yeast. (**A**) Wild-type and *atg32*-null cells expressing mito-DHFR-mCherry were grown under respiratory conditions (Gly), collected at the indicated time points, and subjected to SDS-PAGE and Western blotting. Generation of free mCherry indicates transport of mitochondria to the vacuole. The amounts of free mCherry were quantified at the 24, 36, 48, and 72 h time points. The signal intensity value of free mCherry in wild-type cells at 72 h was set to 100%. Data represent the averages of three experiments, with error bars indicating standard deviations. Por1 and Pgk1 were monitored as an endogenous mitophagy marker and loading control, respectively. (**B**) Wild-type and *atg32*-null cells expressing

free mCherry that is highly resistant against vacuolar proteases. Thus, Atg32-dependent accumulation of free mCherry semiquantitatively indicates the relative level of mitophagy (*see* **Note 8**) [16].

1. Prepare cell cultures under prolonged respiration and starvation as described in Section 3.3.1, **steps 1 and 2**, and Section 3.3.2, **steps 1–5**, respectively.
2. Transfer the cells (1 OD_{600} units) to a 1.5-mL microcentrifuge tube at the 0 (from SDCA), 24, 36, 48, and 72 h time points for mitophagy under prolonged respiration or 0 (from YPD), 6, 12, 24, and 48 h time points for mitophagy under prolonged starvation, harvest them by centrifugation (16,000 *g*, 15 s), discard the supernatant using a micropipette, and store the cell pellet at −80°C before use.
3. Add 100 μL 0.1 *M* NaOH to the cell pellet, vortex vigorously, and keep at room temperature for 5 min.
4. Harvest the cells by centrifugation (16,000 *g*, 2 min), and discard the supernatant using a micropipette.
5. Add 50 μL sample buffer to the cell pellet, and vortex vigorously.
6. Boil the cell suspension for 3 min and keep them on ice for 1 min.
7. Separate the supernatant by centrifugation (16,000 *g*, 2 min), and transfer it to a 1.5-mL microcentrifuge tube.
8. Analyze 5–10 μL samples (0.1–0.2 OD_{600} units) by 10% SDS-PAGE, and blot them to a PVDF membrane.
9. Block the membrane with 5% skim milk in TBS-T at room temperature for 30 min.
10. Incubate the membrane with 0.1% skim milk in TBS-T containing the anti-mCherry (1:2,000 dilution), anti-Por1 (1:1,000 dilution), or anti-Pgk1 (1:10,000 dilution) antibody at room temperature for 1 h (*see* **Note 9**).
11. Wash the membrane three times with TBS-T for 7 min each.
12. Incubate the membrane with 1% skim milk in TBS-T containing HRP-conjugated anti-mouse IgG (1:10,000 dilution) at room temperature for 1 h.
13. Wash the membrane three times with TBS-T for 7 min each.

Fig. 2 (continued) mito-DHFR-mCherry were pregrown under nutrient-rich conditions, shifted to nitrogen starvation (−N), collected at the indicated time points, and subjected to SDS-PAGE and Western blotting. The amounts of free mCherry were quantified at the 6, 12, 24, and 48 h time points. The signal intensity value of free mCherry in wild-type cells at 48 h was set to 100%. Data represent the averages of three experiments, with error bars indicating standard deviations

14. Incubate the membrane with ECL detection regents, and detect protein signals using a luminescent image analyzer (*see* **Note 10**).
15. For free mCherry bands, prepare three Western blots, perform immunodecoration and ECL detection, and quantify protein signals using a luminescent image analyzer.

4 Notes

1. The plasmid pBSII-TEF^{P}-mito-DHFR-mCherry-CgHIS3 contains the pBluescript II backbone (high-copy *E. coli* cloning vector with ampicillin resistance) and the yeast expression cassette encoding *mito-DHFR-mCherry* and *CgHIS3* genes. The mitochondrial matrix-localized marker mito-DHFR-mCherry consists of dihydrofolate reductase (DHFR)-mCherry fused at the C-terminus of the mitochondrial targeting sequence derived from the filamentous fungus *Neurospora crassa* ATP synthase subunit 9 (amino acids 1–69). CgHis3 is an imidazoleglycerol-phosphate dehydratase derived from the yeast pathogen *Candida glabrata* and can complement histidine auxotrophy of the *S. cerevisiae* BY4741 strain. The expression of *mito-DHFR-mCherry* and *CgHIS3* is under the control of the strong constitutive *TEF2* and endogenous *CgHIS3* promoters, respectively.
2. Conventional strains such as W303 and SEY6210 can also be used to generate derivatives for mitophagy assays using the protocols described here. It should, however, be noted that BY4741 grows faster in non-fermentable medium and undergoes mitophagy more drastically than W303 and SEY6210.
3. To perform high-efficiency transformation, it is critical to use cells at mid-log phase.
4. When targeted to the mitochondrial matrix, mito-DHFR-mCherry is processed to become a mature form (DHFR-mCherry). Under a fluorescence microscopy, DHFR-mCherry can visualize elongated tubular structures (typical mitochondrial patterns) at the cell periphery (*see* Fig. 1B). In addition, Western blotting using the anti-mCherry antibody can detect DHFR-mCherry as a 49 kDa band (*see* Fig. 2A, B).
5. The red fluorescent protein mCherry forms a β-barrel structure that is highly resistant against vacuolar proteases and stays fluorescent in the vacuole.

6. To activate the autophagy machinery under nitrogen deprivation, it is necessary to use cells pregrown in rich YPD medium at mid-log phase.
7. If a residual amount of YPD remains in the cell suspension, the autophagy machinery cannot fully be activated. Thus, washing steps with SD-N should carefully be performed.
8. Accumulation of free mCherry depends on its transport to the vacuole and turnover in the vacuole. It is, however, still possible that the vacuolar protease activities are not always constitutive and may vary by growth phase and genetic background.
9. The mouse monoclonal mCherry antibody ([1C51] ab125096, Abcam) is quite specific and does not provide any nonspecific cross-reacting bands for yeast whole cell extracts.
10. The protein expression profiles of Por1, a mitochondrial outer membrane-localized voltage-dependent anion channel (VDAC), and Pgk1, a cytosolic 3-phosphoglycerate kinase for glycolysis and gluconeogenesis, should also be examined as an endogenous mitophagy marker and loading control, respectively [16].

Acknowledgments

We thank Akinori Eiyama for valuable comments on this manuscript, and Noriko Kondo-Okamoto for constructing yeast strains and establishing the original methods. This work was supported by JSPS KAKENHI Grant Number 16H04784 and MEXT KAKENHI Grant Number 16H01203.

References

1. Mishra P, Chan DC (2016) Metabolic regulation of mitochondrial dynamics. J Cell Biol 212:379–387
2. Okamoto K, Kondo-Okamoto N (2012) Mitochondria and autophagy: critical interplay between the two homeostats. Biochim Biophys Acta 1820:595–600
3. Labbe K, Murley A, Nunnari J (2014) Determinants and functions of mitochondrial behavior. Annu Rev Cell Dev Biol 30:357–391
4. Scheibye-Knudsen M, Fang EF, Croteau DL, Wilson DM 3rd, Bohr VA (2015) Protecting the mitochondrial powerhouse. Trends Cell Biol 25:158–170
5. Youle RJ, van der Bliek AM (2012) Mitochondrial fission, fusion, and stress. Science 337:1062–1065
6. Mishra P, Chan DC (2014) Mitochondrial dynamics and inheritance during cell division, development and disease. Nat Rev Mol Cell Biol 15:634–646
7. Okamoto K (2014) Organellophagy: eliminating cellular building blocks via selective autophagy. J Cell Biol 205:435–445
8. Yamano K, Matsuda N, Tanaka K (2016) The ubiquitin signal and autophagy: an orchestrated dance leading to mitochondrial degradation. EMBO Rep 17:300–316
9. Nguyen TN, Padman BS, Lazarou M (2016) Deciphering the molecular signals of PINK1/Parkin mitophagy. Trends Cell Biol 26:733–744
10. Liu L, Sakakibara K, Chen Q, Okamoto K (2014) Receptor-mediated mitophagy in yeast and mammalian systems. Cell Res 24:787–795

11. Wei H, Liu L, Chen Q (2015) Selective removal of mitochondria via mitophagy: distinct pathways for different mitochondrial stresses. Biochim Biophys Acta 1853:2784–2790
12. Murakawa T et al (2015) Bcl-2-like protein 13 is a mammalian Atg32 homologue that mediates mitophagy and mitochondrial fragmentation. Nat Commun 6:7527
13. Kanki T, Wang K, Cao Y, Baba M, Klionsky DJ (2009) Atg32 is a mitochondrial protein that confers selectivity during mitophagy. Dev Cell 17:98–109
14. Okamoto K, Kondo-Okamoto N, Ohsumi Y (2009) Mitochondria-anchored receptor Atg32 mediates degradation of mitochondria via selective autophagy. Dev Cell 17:87–97
15. Eiyama A, Kondo-Okamoto N, Okamoto K (2013) Mitochondrial degradation during starvation is selective and temporally distinct from bulk autophagy in yeast. FEBS Lett 587:1787–1792
16. Sakakibara K et al (2015) Phospholipid methylation controls Atg32-mediated mitophagy and Atg8 recycling. EMBO J 34:2703–2719

Methods in Molecular Biology (2018) 1759: 85–93
DOI 10.1007/7651_2017_12

Published online: 22 March 2017

MitoPho8Δ60 Assay as a Tool to Quantitatively Measure Mitophagy Activity

Zhiyuan Yao, Xu Liu, and Daniel J. Klionsky

Abstract

Mitophagy, a selective type of macroautophagy (hereafter referred to as autophagy), specifically mediates the vacuole/lysosome-dependent degradation of damaged or surplus mitochondria. Because this process regulates the number and quality of mitochondria, it is vital for proper cellular homeostasis. Mitophagy also plays critical roles in the clearance of paternal mitochondria in *C. elegans* embryos, in erythroid cell maturation, and in the prevention of neurodegenerative disease and cancer. In order to study the molecular mechanism and regulation of mitophagy, sensitive assays are necessary to quantitatively measure mitophagy activity. In the budding yeast, *Saccharomyces cerevisiae*, a "mitoPho8Δ60" assay was developed to study mitophagy. In this assay, Pho8, a vacuolar phosphatase protein, is genetically engineered to be targeted to mitochondria. When mitophagy is induced, the phosphatase protein, along with mitochondria, is conveyed to the vacuole, where its C-terminal propeptide is removed and the phosphatase activity becomes activated; under growing conditions only a background level of delivery occurs. For this reason, the enzymatic activity of mitoPho8Δ60 is correlated with the amount of mitochondria delivered to the vacuole. Thus, this assay serves as a very convenient tool to quantitatively monitor mitophagy activity in yeast.

Keywords: Autophagy, Mitochondria, Mitophagy, Vacuole, Stress, Yeast

1 Introduction

Autophagy is a process where cellular components are targeted for degradation in the vacuole in yeasts and plants or the lysosome in mammals; the delivered cargo is typically degraded, and the resulting macromolecules are subsequently recycled after being released back into the cytosol [1–3]. Autophagy can be either nonselective or selective. When it is in a selective mode, autophagy specifically recognizes particular cargoes and targets them for degradation [4]. Examples of selective autophagy pathways include mitophagy (targeting mitochondria), pexophagy (targeting peroxisomes), and the cytoplasm-to-vacuole targeting pathway (a biosynthetic process that delivers certain resident hydrolases to the vacuole) [4, 5]. In yeast, although nonselective and selective autophagy pathways share the same core molecular machinery, selective autophagy pathways require the scaffold protein Atg11 and corresponding receptor proteins that recognize different cargoes to achieve specificity [4]. For instance, Atg32 is the receptor for mitophagy, Atg36 for pexophagy and Atg19 for the cytoplasm-to-vacuole targeting pathway [6–9].

Mitochondria are essential organelles that regulate cellular energy homeostasis. Oxidative phosphorylation involving the mitochondrial electron transport chain generates reactive oxygen species that can be damaging to DNA and other cellular components. Therefore, dysfunctional or superfluous mitochondria, which may generate excessive reactive oxygen species, can be detrimental to the cell, possibly resulting in oncogenic mutations and/or resulting in the initiation of apoptosis [10]. Accordingly, proper quality and quantity control of this organelle is indispensable. Mitophagy plays an important role in this control process, as it conveys damaged or surplus mitochondria to the lysosome/vacuole for degradation. Recent studies have also shown that mitophagy is associated with normal development and differentiation. For example, in *C. elegans* embryos, the sperm-derived paternal mitochondria are eliminated through mitophagy [11], and this process is similarly important during the maturation of red blood cells [12]. Conversely, defective mitophagy is implicated in the onset of certain neurodegenerative diseases [13, 14]. Thus, understanding the mechanism of mitophagy has important implications for various therapeutic approaches.

To study the molecular mechanism of mitophagy and how the process is regulated, it is necessary to develop quantitative assays to monitor mitophagy activity in cells. In the yeast *Saccharomyces cerevisiae*, the mitoPho8Δ60 assay is now commonly used in the field to measure mitophagy activity. This assay takes advantage of Pho8, a resident vacuolar phosphatase. Pho8 is a type II transmembrane protein with the N-terminal tail facing toward the cytosol and the C terminus inside the vacuolar lumen [15]. The precursor form of Pho8, which is kept inactive due to the presence of a C-terminal propeptide, is delivered to the vacuole through a part of the secretory pathway. After vacuolar delivery, the propeptide is removed, generating the active form of the protein [15]. The mitoPho8Δ60 assay utilizes a mutant form of Pho8 in which the N-terminal 60 amino acids have been removed; this part of the protein functions as an internal uncleaved signal sequence, normally allowing translocation into the endoplasmic reticulum. Thus, Pho8Δ60 is unable to enter the secretory pathway and remains in the cytosol. This truncated form of Pho8 is fused to a mitochondrial inner membrane protein, Cox4, to generate mitoPho8Δ60 [16]. The Cox4-Pho8Δ60 fusion protein is targeted to mitochondria after its synthesis and then becomes sequestered by a phagophore under mitophagy-inducing conditions, leading to its delivery to the vacuole and subsequent activation (Fig. 1). Because of the relatively low level of phosphatase activity that occurs under basal conditions,

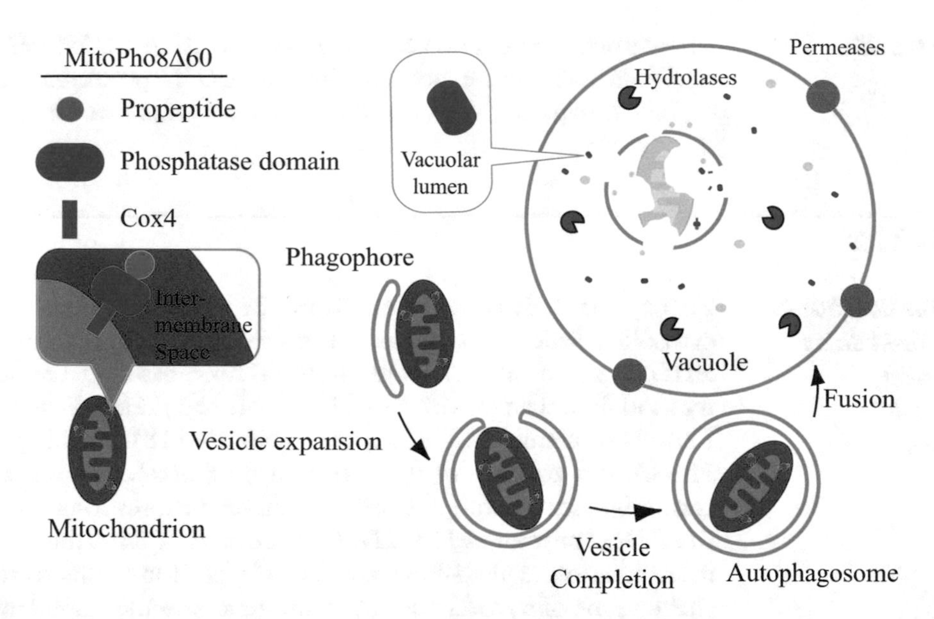

Fig. 1 A schematic model of the mitoPho8Δ60 assay. The mitochondria-targeting sequence in Cox4 allows localization of the Cox4-Pho8Δ60 fusion protein, containing the phosphatase domain and the propeptide of Pho8, onto the mitochondrial inner membrane. Upon induction of mitophagy, the fusion protein as part of a mitochondrion is captured by a phagophore and delivered to the vacuole. In the vacuolar lumen, the propeptide is removed, and the phosphatase activity becomes activated

by measuring the phosphatase activity associated with mitoPho8Δ60, we are able to monitor mitophagy activity in a quantitative manner.

2 Materials

2.1 Culture Medium

1. YPD: 1% yeast extract (ForMedium), 2% peptone (ForMedium), and 2% glucose.
2. YPL: 1% yeast extract, 2% peptone, and 2% lactate (Sigma).
3. SD-N: 0.17% yeast nitrogen base without amino acids and ammonium sulfate (ForMedium) and 2% glucose.

2.2 Buffers

1. Lysis buffer: 20 mM PIPES (Research Organics/Sigma), 0.5% Triton X-100 (Sigma), 50 mM KCl, 100 mM potassium acetate, 10 mM $MgSO_4$, 10 mM $ZnSO_4$, and 1 mM PMSF (Sigma).
2. Reaction buffer: 250 mM Tris-HCl, pH 8.5; 0.4% Triton X-100; 10 mM $MgSO_4$, 10 mM $ZnSO_4$.
3. Stop buffer: 1 M glycine-KOH, pH 11.0.

2.3 Plasmid

1. Mitochondrial Pho8Δ60 plasmid: *ADH1p-COX4-PHO8Δ60 (406)*. This construct contains the *ADH1* promoter and uses the integrating vector pRS406 that contains a uracil selectable marker.

3 Methods

3.1 The Construction of the Yeast Strain Expressing Mitochondrial Pho8Δ60

To exclude activity resulting from the endogenous vacuolar or cytosolic alkaline phosphatases, the wild-type genomic *PHO8* and *PHO13* genes are removed using a loxP-based gene deletion method described previously [17] (*see* **Note 1**). The mitochondrial Pho8Δ60 plasmid was described previously [18]. In this plasmid, Pho8Δ60 is attached to the C terminus of Cox4, which is a mitochondrial inner membrane protein; chimera expression is driven by the *ADH1* promoter. The *ADH1p-COX4-PHO8Δ60* fragment was inserted into the pRS406 vector. The integration of this vector into the genome of a *pho8Δ pho13Δ* strain generates the Pho8Δ60 yeast strain. The detailed integration steps are described below:

1. Obtain the mitochondrial Pho8Δ60 plasmid described previously [18], and prepare plasmid DNA using standard procedures or a commercial kit.
2. Digest 2 μl plasmid with the restriction enzyme StuI in a total volume of 20 μl. Incubate at 37°C for least 3 h (*see* **Note 2**).
3. Add 1 μl of CIP and incubate for an additional 45 min. CIP catalyzes the dephosphorylation of 5′ and 3′ ends of DNA phosphomonoesters, preventing the religation of the plasmid.
4. Transform 5 μl of the digestion product into the *pho8Δ pho13Δ* strain using a standard LiAc method [19]. The cells are spread onto SMD plates lacking uracil to select for integration of the plasmid DNA; incubate at 30°C for 2 days.
5. Pick grown colonies and streak them separately on the uracil selection plate again to further purify individual transformants; incubate the plates at 30°C overnight, then pick the growing strains, restreak them on YPD plates, and incubate at 30°C overnight. The YPD plate can be kept at 4°C for 4–6 weeks.

3.2 Assay for MitoPho8Δ60 Activity

In the mitoPho8Δ60 assay, the procedure used for the measurement of vacuolar phosphatase activity is similar to that in the nonselective Pho8Δ60 autophagy assay. The major difference between the two assays concerns the induction process. For the mitoPho8Δ60 assay, yeast cells first need to be cultured in a nonfermentable carbon source medium for mitochondria proliferation (they can alternatively be grown to a post-logarithmic phase for 1–3 days [7]). After the cells are shifted back to a medium containing a fermentable carbon source, the excess mitochondria will result in cellular stress causing the induction of mitophagy. However, the level of mitochondria-dependent stress induced by this growth regimen alone is not sufficient to create a strong enough signal to allow sensitive detection by the mitoPho8Δ60 activity assay. Therefore, in addition to inducing stress from excess mitochondria, nitrogen starvation is also utilized to increase cell stress and to amplify the phosphatase activity signal. Although nitrogen starvation also induces nonselective autophagy, the Cox4-Pho8Δ60 chimera only resides in mitochondria, so this assay can still provide an accurate reflection of the mitophagy activity. To control for potential background activity, we recommended two controls: First, conduct the nonselective Pho8Δ60 assay in an identical strain (but expressing Pho8Δ60 instead of mitoPho8Δ60) grown in the same conditions. Second, determine mitophagy activity in the mitoPhoΔ60 strain that has been deleted for the *ATG32* gene, which is essential for mitophagy.

1. The strains of interest are grown in 2 ml YPD medium to mid-log phase (O.D.$_{600}$ = 0.8 ~ 1.0). The proper positive and negative control strains are required in each experiment (*see* **Note 3**).
2. Collect the cells by centrifugation at 1,500 × *g*, 2 min, and decant the supernatant. Wash the cells with sterile water or YPL medium one time by vortex followed by centrifugation. Decant the supernatant and resuspend the pellet in YPL medium. Dilute the cells into the proper volume of YPL medium so that the starting O.D.$_{600}$ = 0.1 (*see* **Note 4**). Typically 1.0 O.D.$_{600}$ unit of cells (i.e., 1 ml of cells at an O.D.$_{600}$ = 1.0) is needed for each sample for the assays. For each strain, at least one growing sample and one starvation sample are required. Thus, it is recommended to start the culture with ~3–4 ml of YPL medium. You can increase the volume if more than two time points of samples are required. Culture cells in a 30°C water bath shaker for 12 h until the O.D.$_{600}$ = ~0.8 – 1.0.
3. Harvest the "growing condition" samples. Collect 1.0 O.D.$_{600}$ unit of cells for each strain by centrifugation in a 1.7-ml microcentrifuge tube at 1,500 × *g*, 2 min. Discard the supernatant, and wash the pellets with 1 ml ice cold 0.85% NaCl containing 1 mM PMSF. Collect the cells by centrifugation and discard the supernatant. Store these growing condition samples at −20°C.
4. Harvest the remainder of the cells by centrifugation at 1,500 × *g*, 2 min. Remove the supernatant, wash the pellets with sterile water one time then repeat the centrifugation, and again discard the supernatant. Resuspend the cells in the proper volume of SD-N medium to obtain an O.D.$_{600}$ = ~0.8 – 1.0. Transfer the cells into culture tubes or flasks, and culture cells in the 30°C water bath for 6 h (*see* **Note 5**).

5. Harvest the "starvation" samples. Collect 1.0 O.D.$_{600}$ units of cells for each strain in a 1.7-ml microcentrifuge tube by centrifugation at 1,500 × *g*, 2 min. Discard the supernatant, and wash the pellets with 1 ml cold 0.85% NaCl containing 1 mM PMSF. Place these starvation samples at −20°C until you are ready to proceed (*see* **Note 6**).
6. Lysis of the cells. For each sample, the pellet is resuspended in 200 μl ice cold lysis buffer. Note that the PMSF should be added just before use due to its short half-life in aqueous solution. Add approximately 100 μl acid-washed glass beads [20] into the cell pellet. Vortex the samples vigorously for 5–10 min at 4°C. Centrifuge the samples with maximum speed (16,100 × g) for 5 min at 4°C, and transfer the supernatant into new tubes for subsequent analysis.
7. Prepare the substrate buffer. Add pNPP into the proper volume of reaction buffer to make the final pNPP concentration 1.25 mM (*see* **Note 7**). The tube containing the substrate buffer should be wrapped in aluminum foil or prepared in an opaque bottle to avoid light. Mix the solution by shaker or by gently inverting the tubes. Make enough of the substrate buffer so that you have 400 μl for each sample as well as a blank control. Pre-warm the reaction and substrate buffers at 37°C.
8. Prepare the proper volume of buffer for the BCA assay, which is used to determine the protein concentration. You need 1 ml BCA buffer for each sample and the blank control. Mix the BCA reagent A, BCA reagent B, and water in the ratio of 1:0.02:1. Pre-warm the buffer at 37°C.
9. Transfer 100 μl of the supernatant of each sample from step 6 into a new 1.7-ml tube. Add 100 μl of lysis buffer into a tube as the blank control. Add 400 μl pre-warmed substrate buffer for each sample. Gently invert the tubes several times, and incubate the samples at 37°C for ~10–20 min (record the time; *see* **Note 8**).
10. Stop the reaction using 500 μl of stop buffer. Gently invert the tubes several times. Centrifuge the tubes at maximum speed (16,100 × *g*), 2 min to remove possible precipitates. Measure the O.D.$_{400}$ of the samples, record the readouts, and determine the consumed amount of pNPP using a standard curve (*see* **step 11**).
11. MitoPho8Δ60 activity assay standard curve generation. Pho8Δ60 cleaves p-nitrophenyl phosphate to produce p-nitrophenol. Thus, different concentrations of p-nitrophenol are made for the standard curve. We suggest a standard curve using 1–100 nmoles of p-nitrophenol. These concentrations of p-nitrophenol should result in O.D.$_{400}$ values between ~0 and 2.0. In this range, the relation between the O.D.$_{400}$ value and the concentration of p-nitrophenol

should be linear, although this depends on the spectrophotometer being used. After linear regression, the correlation between O.D.$_{400}$ value and concentration of p-nitrophenol is determined. Based on the O.D.$_{400}$ values obtained in **step 10** and the standard curve, the concentration of p-nitrophenol generated in the samples is determined.

12. Protein concentration determination. Transfer 30 μl lysate of each sample from **step 6** into a new 1.7-ml tube. Place 30 μl of lysis buffer into a tube as the blank control. Add 1 ml BCA buffer into each sample, invert the tubes several times, and incubate the tubes at 37°C for ~20–30 min. Measure the O.D.$_{562}$ of samples, record the readouts, and determine the protein concentration using a standard curve (*see* **step 13**).
13. BCA standard curve generation. A range of BSA from 0 to 1.0 mg/ml is used to generate the standard curve. This should result in O.D.$_{562}$ values from 0 to 1.0. The determination of protein concentration in the samples proceeds as described in the end of **step 11**.

This method will generate reproducible and quantitative results suitable for statistical analysis. After obtaining the generated p-nitrophenol concentration and the total protein concentration, together with the mitoPho8Δ60 reaction time, the mitophagy activity is calculated as nmol p-nitrophenol/min/mg protein.

In addition to providing an absolute value, the mitophagy activity can also be presented as a relative value, in which case the wild-type starvation value is set as 100%, and all other values are normalized to the wild type (Fig. 2) [21]. This type of presentation

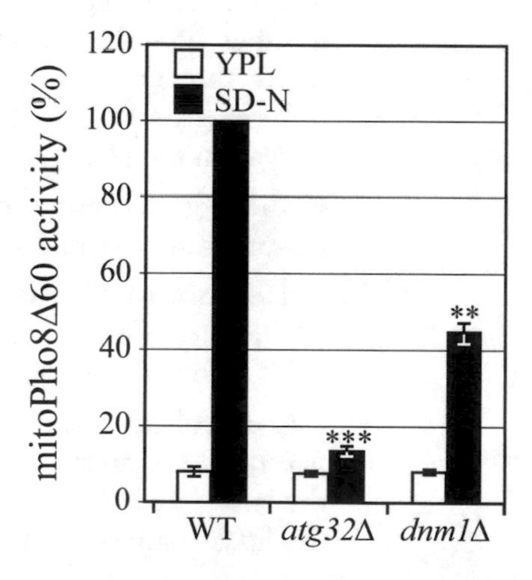

Fig. 2 Relative mitoPho8Δ60 activity (modified from Mao et al. [21] with permission from Elsevier). MitoPho8Δ60 activity was measured in wild-type, *atg32Δ* and *dnm1Δ*, strains after growing in YPL medium to mid-log phase and then shifting to SD-N for 6 h. The wild-type activity was set to 100%, and other strains were normalized

can be used to show the increased or decreased percentage of mitophagy activity in mutant cells compared to the wild type. When using this method of presentation in replicate experiments, it is important to keep the absolute values of the wild-type constant.

4 Notes

1. To preserve auxotrophic markers so the strain can be further modified genetically, a plasmid expressing the Cre enzyme can be transformed into the *pho8Δ pho13Δ* strain to loop out the selection marker gene. The detailed steps can be found in the corresponding reference [17].
2. The digestion product can be stored at −20°C until needed.
3. The positive control for the mitoPho8Δ60 assay is the wild-type mitoPho8Δ60 strain described above. The negative control is a mitoPho8Δ60 strain with deletion of an *ATG* gene that is involved in mitophagy. The *atg32Δ* or *atg11Δ* strain is commonly used as the negative control.
4. The cell growth rate in YPL is much slower than that in YPD. Moreover, the rate varies among different strains. The starting $O.D._{600}$ value of the culture needs to be adjusted based on the empirically determined growth rate in order to get mid-log cells after 12 h incubation in YPL.
5. Cells will grow initially after transfer into the SD-N medium, although at a very slow rate; they will typically undergo a single doubling. Thus, it is recommended to add slightly more medium than required based on the strict calculation for the number of units needed and the appropriate dilution (i.e., dilute the cells appropriately considering that they will continue to grow) and measure the $O.D._{600}$ before harvesting the starvation samples.
6. If there is not enough time for the subsequent assays, all samples can be stored at −20°C until you are ready to proceed.
7. A 100 mM pNPP stock solution can be made by dissolving pNPP into Millipore-filtered water. This stock can be stored at −20°C for a few months.
8. The exact length of time for incubation varies and needs to be determined empirically. The stop time should be a point when the substrate is still saturating, and while the readout is still in the linear range of the standard curve ($O.D._{400} < 2.0$).

Acknowledgments

This work was supported by NIH grant GM053396 to D.J.K. and a Rackham Predoctoral Fellowship to X.L.

References

1. Deretic V, Levine B (2009) Autophagy, immunity, and microbial adaptations. Cell Host Microbe 5:527–549
2. Klionsky DJ, Codogno P (2013) The mechanism and physiological function of macroautophagy. J Innate Immun 5:427–433
3. Xie Z, Klionsky DJ (2007) Autophagosome formation: core machinery and adaptations. Nat Cell Biol 9:1102–1109
4. Johansen T, Lamark T (2011) Selective autophagy mediated by autophagic adapter proteins. Autophagy 7:279–296
5. Jin M, Liu X, Klionsky DJ (2013) SnapShot: selective autophagy. Cell 152:368–368.e2
6. Kanki T, Wang K, Cao Y, Baba M, Klionsky DJ (2009) Atg32 is a mitochondrial protein that confers selectivity during mitophagy. Dev Cell 17:98–109
7. Okamoto K, Kondo-Okamoto N, Ohsumi Y (2009) Mitochondria-anchored receptor Atg32 mediates degradation of mitochondria via selective autophagy. Dev Cell 17:87–97
8. Motley AM, Nuttall JM, Hettema EH (2012) Pex3-anchored Atg36 tags peroxisomes for degradation in *Saccharomyces cerevisiae*. EMBO J 31:2852–2868
9. Scott SV, Guan J, Hutchins MU, Kim J, Klionsky DJ (2001) Cvt19 is a receptor for the cytoplasm-to-vacuole targeting pathway. Mol Cell 7:1131–1141
10. Wallace DC (2005) A mitochondrial paradigm of metabolic and degenerative diseases, aging, and cancer: a dawn for evolutionary medicine. Annu Rev Genet 39:359–407
11. Sato M, Sato K (2011) Degradation of paternal mitochondria by fertilization-triggered autophagy in *C. elegans* embryos. Science 334:1141–1144
12. Sandoval H, Thiagarajan P, Dasgupta SK, Schumacher A, Prchal JT, Chen M, Wang J (2008) Essential role for Nix in autophagic maturation of erythroid cells. Nature 454:232–235
13. Youle RJ, Narendra DP (2011) Mechanisms of mitophagy. Nat Rev Mol Cell Biol 12:9–14
14. Ashrafi G, Schwarz TL (2013) The pathways of mitophagy for quality control and clearance of mitochondria. Cell Death Differ 20:31–42
15. Klionsky DJ, Emr SD (1989) Membrane protein sorting: biosynthesis, transport and processing of yeast vacuolar alkaline phosphatase. EMBO J 8:2241–2250
16. Kanki T, Wang K, Klionsky DJ (2010) A genomic screen for yeast mutants defective in mitophagy. Autophagy 6:278–280
17. Gueldener U, Heinisch J, Koehler GJ, Voss D, Hegemann JH (2002) A second set of loxP marker cassettes for Cre-mediated multiple gene knockouts in budding yeast. Nucleic Acids Res 30:e23
18. Kanki T, Wang K, Baba M, Bartholomew CR, Lynch-Day MA, Du Z, Geng J, Mao K, Yang Z, Yen WL, Klionsky DJ (2009) A genomic screen for yeast mutants defective in selective mitochondria autophagy. Mol Biol Cell 20:4730–4738
19. Gietz RD, Woods RA (2002) Transformation of yeast by lithium acetate/single-stranded carrier DNA/polyethylene glycol method. Methods Enzymol 350:87–96
20. Cheong H, Klionsky DJ (2008) Biochemical methods to monitor autophagy-related processes in yeast. Methods Enzymol 451:1–26
21. Mao K, Wang K, Liu X, Klionsky DJ (2013) The scaffold protein Atg11 recruits fission machinery to drive selective mitochondria degradation by autophagy. Dev Cell 26:9–18

Methods in Molecular Biology (2018) 1759: 95–104
DOI 10.1007/7651_2017_13

Published online: 22 March 2017

Mitophagy in Yeast: A Screen of Mitophagy-Deficient Mutants

Kentaro Furukawa and Tomotake Kanki

Abstract

Mitochondrial autophagy (mitophagy) is a process that selectively degrades mitochondria via autophagy. Recent studies have shown that mitophagy plays an important role in mitochondrial homeostasis by degrading damaged or excess mitochondria. The budding yeast *Saccharomyces cerevisiae* is a powerful model organism that has been employed to study several biological phenomena. Recently, there has been significant progress in the understanding of mitophagy in yeast following the identification of Atg32, a mitochondrial outer membrane receptor protein for mitophagy. In this chapter, we describe protocols to study mitophagy in yeast via a genome-wide screen for mitophagy-deficient mutants using fluorescence microscopy and immunoblotting.

Keywords: Yeast, Mitochondria, Mitophagy, Genome-wide screening, Fluorescence microscopy, Immunoblotting

1 Introduction

Mitochondrial autophagy (mitophagy) is a process that selectively degrades mitochondria via autophagy. Increasing evidence indicates that mitophagy plays important roles in mitochondrial quality and quantity control. Recent studies in mammals have identified a number of mitophagy-related factors such as Nix/BNIP3L, BNIP3, FUNDC1, PINK1-Parkin, optineurin, NDP52, and Bcl2-L-13 [1–4]. However, the molecular mechanisms and regulatory roles of these factors in the process of mitophagy are still not fully understood.

The budding yeast *Saccharomyces cerevisiae* is the simplest eukaryote but is a powerful model for studying several biological phenomena. For instance, yeast has contributed to our current understanding of autophagy, including the identification of more than 40 autophagy-related (*ATG*) genes [5, 6]. Similarly, the molecular mechanisms and physiological roles of mitophagy in yeast have been gradually elucidated following the identification of a mitophagy-specific receptor Atg32 [7–15]. In this chapter, we describe a protocol to identify mitophagy-related genes using gene-deletion mutant libraries.

To efficiently induce mitophagy, it is necessary to (pre-)culture cells in non-fermentable medium containing lactate or glycerol as the sole carbon source. This step is important for proliferating mitochondria. When cells are cultured to a stationary phase or are shifted into nitrogen starvation medium, mitophagy is induced to eliminate the proliferated mitochondria. To monitor mitophagy under these conditions, techniques using fluorescence microscopy and/or immunoblotting are commonly used [7, 8]. Additionally, alkaline phosphatase targeted to the mitochondrial matrix (mtALP) and a mitochondrial-targeting sequence fused to the pH-biosensor Rosella (mtRosella) are useful tools for quantitative mitophagy assays [16, 17].

Tagging the C-terminus of the mitochondrial outer membrane protein, Om45 (Om45-GFP), with green fluorescent protein (GFP) makes it possible to visualize mitochondria using green fluorescence [7] (*see* **Note 1**). When mitophagy is induced under the conditions described above, mitochondria carrying Om45-GFP are transported into the vacuole. The mitochondria themselves are immediately degraded by vacuolar hydrolases, while the relatively stable GFP moiety, which is highly resistant to degradation, accumulates in the vacuole. Accordingly, GFP fluorescence in the vacuole can be detected using fluorescence microscopy. When this Om45-GFP system (or some similar system with different fluorescent proteins) is systematically introduced into the yeast gene knockout library, a genome-wide screen for mitophagy-deficient mutants can be performed [8, 17, 18].

2 Materials

2.1 Culture Media and Plate

1. Glucose growth medium, YPD: 1% yeast extract, 2% peptone, and 2% glucose.
2. Synthetic minimal medium with glucose, SMD: 0.67% yeast nitrogen base and 2% glucose, auxotrophic amino acids, and nucleotide bases.
3. Lactate growth medium, YPL: 1% yeast extract, 2% peptone, and 2% lactic acid (pH 5.5).
4. Synthetic minimal medium with glucose lacking nitrogen, SD-N: 0.17% yeast nitrogen base without amino acids and ammonium sulfate, containing 2% glucose.
5. To make agar plates of the above media, add 2% agar to the desired medium (final concentration).

2.2 Solutions

1. 10 × TE solution: 100 mM Tris-HCl, pH 7.5, and 10 mM EDTA, sterilized by filtration.
2. 1 M lithium acetate (pH 7.5 adjusted by acetic acid), sterilized by filtration.
3. TE-lithium acetate: 10 ml of 10 × TE, 10 ml of 1 M lithium acetate, and 80 ml of sterilized water.
4. Single-stranded salmon sperm DNA: 10 mg/ml single-stranded salmon sperm DNA in water, boil for 5 min and then cool on ice.
5. 50% PEG solution: 50% polyethylene glycol 3,350 in water, sterilized by autoclaving.
6. 40% PEG-TE-Li solution: 80 ml of 50% PEG, 10 ml of 10 × TE, and 10 ml of 1 M lithium acetate.
7. Cell lysis buffer: 10 mM Tris-HCl (pH 8.0), 1 mM EDTA, 100 mM NaCl, 1% SDS, and 2% Triton X-100.
8. Phosphate-buffered saline (PBS): 137 mM NaCl, 2.7 mM KCl, 10 mM Na_2HPO_4, 1.8 mM KH_2PO_4, and pH 7.4.
9. Sample buffer for sodium dodecyl sulfate-polyacrylamide gel electrophoresis (SDS-PAGE): 50 mM Tris-HCl (pH 6.8), 2% SDS, 10% glycerol, 5% 2-mercaptoethanol, 1 mM phenylmethylsulfonyl fluoride, and 0.1% bromophenol blue.
10. Washing buffer (PBST): 0.05% Tween-20 in PBS.
11. Blocking solution: 5% nonfat dried milk in PBST.
12. Antibody-binding buffer: 2% nonfat dried milk in PBST.

2.3 Plasmid

1. pFA6a-GFP(S65T)-HIS3MX6 [19]: For adding a C-terminal GFP tag with a selective *HIS3* marker (*see* **Note 2**).

2.4 Yeast Strains

1. BY4742 knockout library [20]: Containing 5,100 mutant strains stocked in 56 96-well plates. Genotype of the host strain, *MATα his3Δ1 leu2Δ0 lys2Δ0 ura3Δ0*.

2.5 Antibodies and Materials

1. Anti-GFP antibody (Clontech, JL-8).
2. Secondary antibody: Horseradish peroxidase (HRP)-conjugated anti-mouse IgG (Millipore, AP124P).
3. Polyvinylidene fluoride (PVDF) membrane (Millipore, IPVH00010).
4. Enhanced chemiluminescence (ECL) detection reagents (Pierce, 32106).
5. Glass beads (Thomas Scientific, 5663R50).
6. Replica plater (SIGMA, R2508).
7. 96-deep-well plate (Thermo Scientific, 260252).
8. 96-well glass bottom plate (IWAKI, 5866-096).

3 Methods

3.1 Genome-Wide Screen for Mitophagy-Deficient Mutants by Fluorescence Microscopy: Initial Screen

3.1.1 Construction of an Om45-GFP-Expressing Strain

1. To chromosomally tag GFP at the C-terminus of Om45 (*see* **Notes 3 and 4**), use the PCR-based gene modification method described by Longtine et al. [19], detailed below.
2. Amplify a DNA fragment encoding GFP and a selective *HIS3* marker via PCR using pFA6a-GFP(S65T)-HIS3MX6 as a template plasmid and the following primers (5′-TGA TAA GGG TGA TGG TAA ATT CTG GAG CTC GAA AAA GGA CCG GAT CCC CGG GTT AAT TAA-3′ and 5′-GAG AAA CAT GTG AAT ATG TAT ATA TGT TAT GCG GGA ACC AGA ATT CGA GCT CGT TTA AAC-3′).
3. Culture BY4742 wild-type cells in a test tube containing 2 ml of YPD medium overnight.
4. Dilute cells from overnight culture in 2 ml of fresh YPD medium and grow to early-log phase ($OD_{600} = 0.8$).
5. Transfer 1.5 ml of cells to a microcentrifuge tube and collect cells by centrifugation at 2,400 × *g* for 20 s, and then aspirate the supernatant.
6. Wash cell pellets by resuspending in 1 ml of sterilized water, centrifuge at 2,400 × *g* for 20 s, and then aspirate the supernatant.
7. Resuspend the cells in 25 μl of TE-lithium acetate. Add 2.5 μl of single-stranded salmon sperm DNA, 150 μl of 40% PEG-TE-Li, and 20 μl of the above Om45-GFP::HIS3MX6 PCR product, and mix well by pipetting.
8. Incubate the cells at 30°C with agitation for 30 min and then at 42°C for 15 min.
9. Centrifuge the cells at 2,400 × *g* for 20 s and then aspirate the supernatant.
10. Resuspend the cells in 100 μl of sterilized water and spread on agar plates containing selective medium (SMD without histidine).
11. Incubate the plate at 30°C for 2–3 days, allowing colonies to form. If GFP is successfully tagged onto the C-terminus of Om45, a typical tubular pattern of mitochondria will be observed by fluorescence microscopy.

3.1.2 Amplification of an Om45-GFP::HIS3 DNA Cassette with Long Flanking Regions

1. Culture the Om45-GFP-expressing wild-type cells overnight in 2 ml of YPD.
2. Transfer cells to a microcentrifuge tube and collect them by centrifugation at 15,000 × *g* for 1 min, and then aspirate the supernatant.
3. Resuspend cells in 200 μl of cell lysis buffer.

4. Add 0.5 g of acid-washed glass beads and 200 μl of Phenol: Chloroform:Isoamyl Alcohol (25:24:1, v/v), and disrupt cells by vortex for 3 min.
5. Centrifuge at 20,000 × *g* for 5 min and transfer aqueous phase (top layer) to a new microcentrifuge tube.
6. Add 500 μl of 99.5% ethanol and centrifuge at 20,000 × *g* for 5 min. A small pellet containing genomic DNA and RNA should be formed. Aspirate the supernatant.
7. Add 500 μl of 80% ethanol and centrifuge at 20,000 × *g* for 5 min. Aspirate the supernatant.
8. Allow pellet to air-dry and then dissolve the pellet in 100 μl of TE with RNase A (20 μg/ml final concentration).
9. Amplify a DNA fragment encoding GFP and a selective *HIS3* marker via PCR, using the above genomic DNA as a template, with the following primers: 5′-GGT GAT ACG GCA CAG GAG TT-3′ and 5′-GTC ACA ACT GGC ACA ACC AC-3′. This PCR product contains long flanking sequences which increase homologous recombination efficiency at the genomic *Om45* locus.

3.1.3 Construction of Gene Knockout Strain Library Expressing Om45-GFP

1. Inoculate yeast knockout strains from originally stored 96-well plates to YPD plate using a replica plater and grow them at 30°C.
2. Re-inoculate cells in 96-deep-well plates containing 400 μl of YPD liquid medium and culture overnight at 30°C.
3. Dilute overnight culture cells in 400 μl of fresh medium and grow to early-log phase ($OD_{600} = 0.8$).
4. Centrifuge the cells at 2,400 × *g* for 2 min and aspirate the supernatant (eight-channel aspirator).
5. Wash the cell pellets by resuspending in 400 μl of sterilized water, centrifuge at 2,400 × *g* for 2 min, and aspirate the supernatant.
6. Resuspend the cells in transformation solution (5 μl of TE-lithium acetate, 0.5 μl of single-stranded salmon sperm DNA, 30 μl of 40% PEG-TE-Li, and 1 μl of Om45-GFP::HIS3MX6 PCR product produced at Section 3.1.2), and mix well by pipetting.
7. Incubate the cells at 30°C for 30 min and then at 42°C for 15 min.
8. Centrifuge the cells at 2,400 × *g* for 2 min and aspirate the supernatant.
9. Resuspend the cells in 50 μl of sterilized water and spot 5 μl of each on agar plates (SMD without histidine) and incubate at

30°C for 2–3 days. Typically, five to ten colonies grow from each spot.

10. Take a small portion of each colony from each spot, suspend in 2 μl of water on the microscope slide, and observe by fluorescence microscopy with a 100× objective. If GFP is successfully tagged onto the C-terminus of Om45, a typical tubular pattern of mitochondria will be observed. Restreak the colonies expressing Om45-GFP on agar plates (SMD without histidine), and store for mitophagy screen.

3.1.4 Induction and Observation of Mitophagy at Stationary Phase by Fluorescence Microscopy

1. Inoculate knockout library strains carrying Om45-GFP on agar plates (SMD without histidine) using a replica plater.
2. Re-inoculate strains in 96-deep-well plates containing 400 μl of YPL liquid medium and culture for up to 72 h.
3. At stationary phase (48–72 h), take 70-μl aliquots of cells and place into 96-well glass bottom plates.
4. Observe GFP fluorescence by fluorescence microscope and determine whether mitophagy occurred or not. When mitophagy occurs, the GFP signal accumulates in the vacuole (Fig. 1 and *see* **Note 5**). Mitophagy-defective strains, which show weak or no accumulation of GFP in the vacuole (*see* **Notes 6 and 7**), are corrected and used for secondary screens (see following step 3.2).

3.2 Secondary Screen for Mitophagy-Deficient Mutants

3.2.1 Induction of Mitophagy by Nitrogen Starvation

1. Culture Om45-GFP-expressing strains selected by the above initial screen in 3 ml of YPD overnight.
2. Dilute overnight culture cells in 3 ml of fresh YPD and grow to early-log phase ($OD_{600} = 0.8$).
3. Centrifuge the cells at 2,400 × *g* for 2 min and aspirate the supernatant.
4. Resuspend the cells in 3 ml of YPL and dilute further with 3 ml of YPL ($OD_{600} = 0.2$). Culture for 12–20 h to allow the cells to grow to log phase again ($OD_{600} = 1.0$–2.0).
5. When the cells have reached log phase, place aliquots equivalent to 1.0 OD_{600} unit in microcentrifuge tubes before making mitophagy induction samples for SDS-PAGE (immediately follow with step 3.2.2.1).
6. Wash the remaining cells twice with 5 ml of sterilized water (centrifuge at 2,400 × *g* for 2 min, aspirate the supernatant, and add sterilized water). After the second washing step, resuspend the cells in 2 ml of SD-N liquid medium and culture for 6 h.
7. After starvation, place aliquots equivalent to 1.0 OD_{600} unit in microcentrifuge tubes to make mitophagy induction samples for SDS-PAGE (immediately follow with step 3.2.2.1).

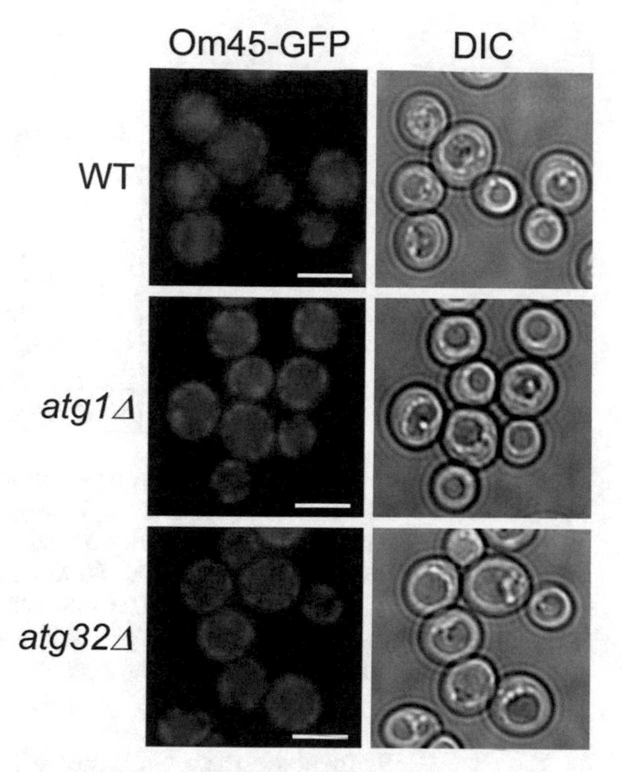

Fig. 1 Monitoring mitophagy by fluorescence microscopy using the mitochondrial protein Om45-GFP at the stationary phase. Wild-type (WT), *atg1Δ*, and *atg32Δ* strains expressing Om45-GFP were cultured in YPL medium for 48 h. The localization of GFP was visualized by fluorescence microscopy. *DIC* differential interference contrast. Scale bars, 5 μm

3.2.2 Detection of Mitophagy by Immunoblotting

1. Add 100% trichloroacetic acid (TCA; 10% final concentration) to the samples prepared above and place the samples on ice for 30 min.
2. Pellet the proteins by centrifugation at 20,000 × *g* for 10 min and aspirate the supernatant.
3. Wash the pellet with 1 ml of ice-cold acetone and allow pellet to air-dry.
4. Resuspend the air-dried pellets in 50 μl of sample buffer with an equal volume of acid-washed glass beads and disrupt by vortexing at 4°C for 3 min.
5. Incubate the samples at 100°C for 5 min.
6. Load 10 μl of each sample (equivalent to 0.2 OD_{600} unit) onto a 12% polyacrylamide gel and subject to electrophoresis.
7. Electrotransfer proteins to a PVDF membrane following a standard semidry Western blotting procedure.
8. Block the PVDF membrane with blocking solution for 30 min with agitation.

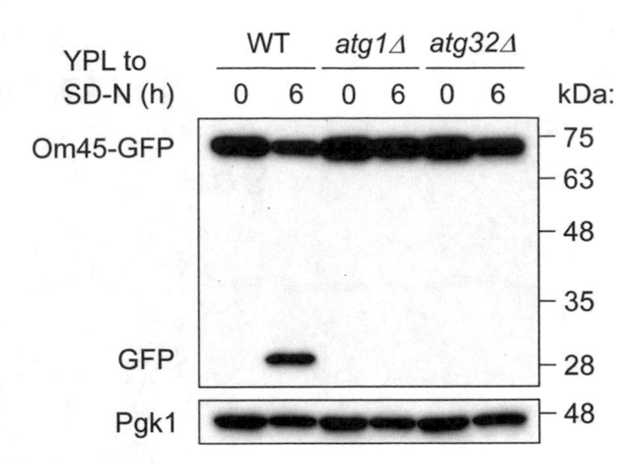

Fig. 2 Monitoring mitophagy by immunoblotting using the mitochondrial protein Om45-GFP during starvation. Wild-type (WT), *atg1Δ*, and *atg32Δ* strains expressing Om45-GFP were cultured in YPL medium to the mid-log growth phase and then shifted to SD-N medium for 6 h. GFP processing was monitored by immunoblotting with anti-GFP antibody. Pgk1 (loading control) was detected using anti-Pgk1 antibody. The positions of molecular weight markers are indicated on the *right*

9. Incubate the membrane with anti-GFP antibody (1:10,000 dilution in antibody-binding buffer) at 4°C overnight.
10. Wash the membrane three times, for 5 min each, in PBST.
11. Incubate the membrane with HRP-conjugated secondary antibody (1:10,000 dilution in antibody-binding buffer) at room temperature for 1 h.
12. Wash the membrane three times, for 10 min each, in PBST.
13. Incubate the membrane with ECL detection reagents, and detect the GFP signals by ChemiDoc XRS Plus (Bio-Rad). Om45-GFP and processed GFP can be detected as approximately 72- and 28-kDa bands, respectively (Fig. 2) (*see* **Note 8**).
14. The signal intensity of Om45-GFP and processed GFP is quantified using Image Lab software to calculate the GFP/Om45-GFP ratio. If there is no processed GFP, mitophagy in the mutant cells is completely inhibited. If GFP/Om45-GFP ratio is lower than that in WT cells, mitophagy in the mutant cells is partially impaired.

4 Notes

1. Although we focus on Om45-GFP in this protocol, isocitrate dehydrogenase (Idh1)-GFP is an alternative construct that has been used successfully [7]. The methods for induction and detection of mitophagy are the same as for Om45-GFP.

2. Plasmids containing other selection markers (pFA6a-GFP (S65T)-natMX or pYM25) are also available [21, 22]. In that case, different selective medium are required to select the positive strains (YPD with 200 μg/ml nourseothricin or with 300 μg/ml hygromycin B, respectively).
3. One important point is that Om45-GFP should not be overexpressed, which can lead to mislocalization and mitophagy-independent degradation. Thus, it is recommended to use the endogenous promoter.
4. Alternatively, an Om45-GFP tagged strain can be purchased form Thermo Fisher Scientific (Yeast GFP Clone Collection).
5. The vacuole membrane can also be stained with the red dye FM 4-64 (Molecular Probes/Invitrogen, T-3166).
6. Most of the *ATG* genes (such as *ATG1–12*, *14*, *16*, *18*, and *32*) are required for mitophagy. Therefore, their deletion mutant strains work as mitophagy-deficient controls.
7. Some knockout strains may not grow well in YPL medium. In most cases, such strains have defects in mitochondrial function, and it is difficult to perform the above mitophagy assay. Similarly, some strains do not express sufficient Om45-GFP for detection of mitophagy. The mtALP assay may be an alternative approach for these strains [16].
8. Anti-Pgk1 antibody (Invitrogen, 459250) is routinely used to detect Pgk1 (3-phosphoglycerate kinase) as a loading control.

Acknowledgments

This work was supported in part by the Japan Society for the Promotion of Science KAKENHI Grant numbers 26291039 (TK), 16H01198 (TK), 16H01384 (TK), 15H06223 (KF), and 16K18514 (KF), Yujin Memorial Grant (Niigata University School of Medicine) (TK), The Sumitomo Foundation (TK), Astellas Foundation for Research on Metabolic Disorders (TK), and Takeda Science Foundation (TK, KF).

References

1. Schweers RL, Zhang J, Randall MS, Loyd MR, Li W, Dorsey FC, Kundu M, Opferman JT, Cleveland JL, Miller JL, Ney PA (2007) NIX is required for programmed mitochondrial clearance during reticulocyte maturation. Proc Natl Acad Sci U S A 104:19500–19505
2. Wu W, Tian W, Hu Z, Chen G, Huang L, Li W, Zhang X, Xue P, Zhou C, Liu L, Zhu Y, Zhang X, Li L, Zhang L, Sui S, Zhao B, Feng D (2014) ULK1 translocates to mitochondria and phosphorylates FUNDC1 to regulate mitophagy. EMBO Rep 15:566–575
3. Lazarou M, Sliter DA, Kane LA, Sarraf SA, Wang C, Burman JL, Sideris DP, Fogel AI, Youle RJ (2015) The ubiquitin kinase PINK1 recruits autophagy receptors to induce mitophagy. Nature 524:309–314

4. Murakawa T, Yamaguchi O, Hashimoto A, Hikoso S, Takeda T, Oka T, Yasui H, Ueda H, Akazawa Y, Nakayama H, Taneike M, Misaka T, Omiya S, Shah AM, Yamamoto A, Nishida K, Ohsumi Y, Okamoto K, Sakata Y, Otsu K (2015) Bcl-2-like protein 13 is a mammalian Atg32 homologue that mediates mitophagy and mitochondrial fragmentation. Nat Commun 6:7527
5. Nakatogawa H, Suzuki K, Kamada Y, Ohsumi Y (2009) Dynamics and diversity in autophagy mechanisms: lessons from yeast. Nat Rev Mol Cell Biol 10:458–467
6. Reggiori F, Klionsky DJ (2013) Autophagic processes in yeast: mechanism, machinery and regulation. Genetics 194:341–361
7. Kanki T, Klionsky DJ (2008) Mitophagy in yeast occurs through a selective mechanism. J Biol Chem 283:32386–32393
8. Okamoto K, Kondo-Okamoto N, Ohsumi Y (2009) Mitochondria-anchored receptor Atg32 mediates degradation of mitochondria via selective autophagy. Dev Cell 17:87–97
9. Kanki T, Wang K, Cao Y, Baba M, Klionsky DJ (2009) Atg32 is a mitochondrial protein that confers selectivity during mitophagy. Dev Cell 17:98–109
10. Mao K, Wang K, Zhao M, Xu T, Klionsky DJ (2011) Two MAPK-signaling pathways are required for mitophagy in *Saccharomyces cerevisiae*. J Cell Biol 193:755–767
11. Aoki Y, Kanki T, Hirota Y, Kurihara Y, Saigusa T, Uchiumi T, Kang D (2011) Phosphorylation of serine 114 on Atg32 mediates mitophagy. Mol Biol Cell 22:3206–3217
12. Kurihara Y, Kanki T, Aoki Y, Hirota Y, Saigusa T, Uchiumi T, Kang D (2012) Mitophagy plays an essential role in reducing mitochondrial production of reactive oxygen species and mutation of mitochondrial DNA by maintaining mitochondrial quantity and quality in yeast. J Biol Chem 287:3265–3272
13. Kanki T, Kurihara Y, Jin X, Goda T, Ono Y, Aihara M, Hirota Y, Saigusa T, Aoki Y, Uchiumi T, Kang D (2013) Casein kinase 2 is essential for mitophagy. EMBO Rep 14:788–794
14. Aihara M, Jin X, Kurihara Y, Yoshida Y, Matsushima Y, Oku M, Hirota Y, Saigusa T, Aoki Y, Uchiumi T, Yamamoto T, Sakai Y, Kang D, Kanki T (2014) Tor and the Sin3-Rpd3 complex regulate expression of the mitophagy receptor protein Atg32 in yeast. J Cell Sci 127:3184–3196
15. Kanki T, Furukawa K, Yamashita S (2015) Mitophagy in yeast: molecular mechanisms and physiological role. Biochim Biophys Acta 1853:2756–2765
16. Müller M, Kötter P, Behrendt C, Walter E, Scheckhuber CQ, Entian KD, Reichert AS (2015) Synthetic quantitative array technology identifies the Ubp3-Bre5 deubiquitinase complex as a negative regulator of mitophagy. Cell Rep 10:1215–1225
17. Böckler S, Westermann B (2014) Mitochondrial ER contacts are crucial for mitophagy in yeast. Dev Cell 28:450–458
18. Kanki T, Wang K, Baba M, Bartholomew CR, Lynch-Day MA, Du Z, Geng J, Mao K, Yang Z, Yen WL, Klionsky DJ (2009) A genomic screen for yeast mutants defective in selective mitochondria autophagy. Mol Biol Cell 20:4730–4738
19. Longtine MS, McKenzie A 3rd, Demarini DJ, Shah NG, Wach A, Brachat A, Philippsen P, Pringle JR (1998) Additional modules for versatile and economical PCR-based gene deletion and modification in *Saccharomyces cerevisiae*. Yeast 14:953–961
20. Giaever G, Chu AM, Ni L, Connelly C, Riles L, Véronneau S, Dow S, Lucau-Danila A, Anderson K, André B, Arkin AP, Astromoff A, El-Bakkoury M, Bangham R, Benito R, Brachat S, Campanaro S, Curtiss M, Davis K, Deutschbauer A, Entian KD, Flaherty P, Foury F, Garfinkel DJ, Gerstein M, Gotte D, Güldener U, Hegemann JH, Hempel S, Herman Z, Jaramillo DF, Kelly DE, Kelly SL, Kötter P, LaBonte D, Lamb DC, Lan N, Liang H, Liao H, Liu L, Luo C, Lussier M, Mao R, Menard P, Ooi SL, Revuelta JL, Roberts CJ, Rose M, Ross-Macdonald P, Scherens B, Schimmack G, Shafer B, Shoemaker DD, Sookhai-Mahadeo S, Storms RK, Strathern JN, Valle G, Voet M, Volckaert G, Wang CY, Ward TR, Wilhelmy J, Winzeler EA, Yang Y, Yen G, Youngman E, Yu K, Bussey H, Boeke JD, Snyder M, Philippsen P, Davis RW, Johnston M (2002) Functional profiling of the *Saccharomyces cerevisiae* genome. Nature 418:387–391
21. Van Driessche B, Tafforeau L, Hentges P, Carr AM, Vandenhaute J (2005) Additional vectors for PCR-based gene tagging in *Saccharomyces cerevisiae* and *Schizosaccharomyces pombe* using nourseothricin resistance. Yeast 22:1061–1068
22. Janke C, Magiera MM, Rathfelder N, Taxis C, Reber S, Maekawa H, Moreno-Borchart A, Doenges G, Schwob E, Schiebel E, Knop M (2004) A versatile toolbox for PCR-based tagging of yeast genes: new fluorescent proteins, more markers and promoter substitution cassettes. Yeast 21:947–962

Methods in Molecular Biology (2018) 1759: 105–110
DOI 10.1007/7651_2017_14

Published online: 22 March 2017

Flow Cytometer Monitoring of Bnip3- and Bnip3L/Nix-Dependent Mitophagy

Matilda Šprung, Ivan Dikic, and Ivana Novak

Abstract

Mitochondria are organelles with numerous vital roles in cellular metabolism. Impaired or damaged mitochondria are degraded in autophagolysosomes in a process known as mitophagy. Given the fundamental role of mitophagy in maintenance of cellular homeostasis, methods and techniques with which to study this process are constantly evolving and emerging. So far, mitophagy flux was mostly monitored using fluorescently labeled LC3 protein on autophagosomal membrane and any of the labeled outer mitochondrial membrane proteins. However, this method is labor intensive, time consuming, and difficult to quantitatively validate due to the rapid mitochondrial turnover. Here, we describe a flow cytometry as a novel and promising quantitative method to monitor Bnip3- and Bnip3L/Nix-mediated mitophagy.

Keywords: Mitophagy, Flow cytometry, Bnip3L/Nix, Mitophagy receptor, MitoTracker

1 Introduction

Mitochondria are double-membrane cell organelles with central role in cellular respiration. Some of the vital metabolic processes such as oxidative phosphorylation, oxidation of fatty acids, and gluconeogenesis take place in these organelles. Other additional mitochondrial cellular functions are ATP production, calcium signaling, production of reactive oxygen species (ROS), and apoptosis [1]. Given the metabolic role, accumulation of dysfunctional or damaged mitochondria can lead to serious pathological conditions including cancer and neurodegenerative disorders. Therefore, mitochondrial quality control mechanisms are of special importance for maintenance of healthy mitochondrial population [2].

Mitophagy is a highly regulated autophagy process during which damaged mitochondria are degraded and removed from the cell. Two distinctive mitophagy pathways are described to date. One involves ubiquitination of OMM proteins via PINK1/Parkin-mediated pathway. Ubiquitinated proteins subsequently recruit autophagosomal membrane through specific receptors that are both able to recognize ubiquitin chains on mitochondrial proteins and LC3 at autophagosomal membrane [3]. Several such receptors have been identified to date, but recently NDP52 and optineurin have been described as primary receptors for PINK1/Parkin-mediated mitophagy [4]. Unlike PINK1/Parkin pathway, which includes ubiquitination as an intermediate step, some mitophagy receptors establish direct interaction with LC3 proteins on autophagosomal membrane and their LC3-interacting region (LIR) situated at N-terminal part of a protein. Four such receptors have been investigated to date, Bnip3, Bnip3L/Nix, Bcl2L13, and FUNDC1 [5–8]. It was shown that Bnip3L/Nix and Bnip3 interact with Atg8 proteins in LIR-dependent manner and recruit autophagic machinery to damaged mitochondria. Moreover, Bnip3L/Nix mediates mitochondrial clearance during reticulocyte differentiation [5], but the detailed mechanisms of its regulation are still to be investigated.

Due to the role that mitophagy has in maintaining cellular homeostasis, it is of a fundamental importance to develop appropriate techniques to study this cell essential process. There are several available techniques to study mitophagy flow in mammalian cells, but mostly mitophagy is monitored by fluorescent microscopy of mitochondrial-autophagosomal colocalization. This approach is labor intensive, time consuming, and quantitatively difficult to validate, and unless lysosomal inhibitors are used, mitochondrial turnover occurs very rapidly making the detection of colocalization events very difficult to spot [9]. Recently, confocal scanning microscopy, transmission electron microscopy, and flow cytometry are emerging as novel methods in mitophagy field [10]. Here, we focus on flow cytometry as rapid, reliable, and quantitative method for detection of Bnip3- and Bnip3L/Nix-mediated mitophagy. Although flow cytometer is still quite expensive instrument, it is often available for the researchers and is, therefore, an excellent and rapid tool to access mitophagy flux. By simple usage of different MitoTracker probes (MitoTracker Green or MitoTracker Red, respectively), it is possible to distinguish between total mitochondrial population and stress-affected one. Further, for more specific analysis, fluorescently tagged mitochondrial and autophagy proteins can be used, like RFP-Bnip3 or Bnip3L/Nix and GFP-LC3, where single- and double-positive subpopulations are analyzed.

To analyze Bnip3- and Bnip3L/Nix-dependent mitophagy flux by flow cytometry, cells can originate from many different sources, from cultured cells to cells isolated from the blood [10, 11]. Here, we will in detail describe how to perform flow cytometry to monitor Bnip3- and Bnip3L/Nix-dependent mitophagy flux. First, we will briefly describe cell culture propagation, and then we will focus on cell and mitochondrial staining procedures as steps precede to flow cytometry. We will conclude by giving the readers information on possible changes in the protocol and alternative approaches.

2 Materials

2.1 Fluorescent Dyes

1. Live/Dead Fixable Violet Dead Cell Stain Kit (*Thermo Fisher Scientific*).
2. MitoTracker Green FM (*Thermo Fisher Scientific*): 1 mM stock solution is prepared in DMSO. Store at −20 °C protected from light. Prepare 150 nM working concentration by diluting in PBS buffer supplemented with 5% FBS.
3. MitoTracker Red CMXRos (*Thermo Fisher Scientific*): 1 mM stock solution is prepared in DMSO. Store at −20 °C protected from light. Prepare 200 nM working concentration by diluting in PBS buffer supplemented with 5% FBS (Table 1).

2.2 Cells, Common Buffers, and Media for Cell Culture

2.2.1 Cells

1. HeLa cells.
2. HEK293 cells.

2.2.2 Cell Culture Reagents and Buffers

1. DMEM (Dulbecco's Modified Eagle Medium) with high glucose, L-glutamine supplemented with 5% FBS (fetal bovine serum) and penicillin/streptomycin solution.
2. Trypsin.
3. Phosphate buffer saline (PBS).
4. Transfection Reagent Kit (any recommendation: jetPRIME transfection reaction kit, *Polyplus*).

2.2.3 Buffers and Reagents for Flow Cytometry

1. DMSO, high molecular grade.
2. 5% FBS (fetal bovine serum): 5% FBS resuspended in phosphate buffer saline (PBS).
3. FACS (fluorescence-activated cell sorting): 2% FBS, 0.02% sodium azide (NaN_3) in PBS. Filter through 0.22 μm pore size filters and store at 4 °C.
4. 2–4% paraformaldehyde. Prepare fresh from paraformaldehyde powder. Filter through 0.22 μm pore size filters to avoid non-dissolved crystals.

Table 1
Properties of the mitochondrial fluorescent dyes

Fluorescent dye	MW (g/mol)	λ_{ex}/nm	λ_{em}/nm	Fixable
MitoTracker Red CMXRos	532	578	599	Yes
MitoTracker Green FM	672	490	516	No

3 Methods

In order to monitor mitophagy flux in cell lines such as HEK293 or HeLa by flow cytometry, only population of viable cells needs to be taken into account. Discrimination of live vs. dead cells is achieved by cell-viability staining applying Live/Dead Fixable Violet Dead Cell Stain Kit. When the cell membrane is disrupted, the dye freely defuses in the cell interior resulting in an intense fluorescence staining. In viable cells, only surface of the cell is stained, and the difference in the fluorescence intensity between live and dead cells is typically greater than 50-fold allowing easy and efficient discrimination of the two cell populations.

For mitochondrial staining, two fluorescent dyes could be used, MitoTracker Red CMXRos (MTR) and/or MitoTracker Green FM (MTG). MTR is used for detection of depolarized mitochondria, whereby MTG is less specific and stains both polarized and depolarized mitochondria. To distinguish between the two different mitochondrial populations, both dyes should simultaneously be used.

3.1 Cell Preparation

1. Seed 0.5–1.25 × 10^5 cells/well and culture cells (HEK293, HeLa) in a complete DMEM containing 5% FBS and penicillin/streptomycin. Incubate cells in a humidified tissue culture incubator with 5% CO_2, at 37 °C for 24 h until they reach 60–70% confluency.
2. Transfect cells with transfection reagent of choice using GFP- or RFP-Bnip3L/Nix according to manufacturer's instructions (recommendation to use jetPRIME transfection reaction kit, *Polyplus*) (*see* **Notes 1–3** for alternative). Incubate cells for another 24 h in a humidified tissue culture incubator with 5% CO_2, at 37 °C.
3. If using adherent cells, trypsinize first to achieve single cell population and avoid cell clumping. Also filter cells through 70 μm Cell Strainer (to avoid cell clumps that might clog flow cytometer). Transfer ≤1 × 10^6 cells to microcentrifuge tubes.
4. Centrifuge cells at 500 × *g* for 5 min at 4 °C and discard supernatant.
5. Add 100 μL of cell-viability Live/Dead Fixable Violet Dead Cell Stain at a dilution of 1:1,000 in PBS. Gently resuspend cells and incubate at room temperature for 10 min while protecting them from light.
6. Wash the cells twice with 200 μL of pre-chilled FACS buffer. Pellet cells at 500 × *g* for 5 min at 4 °C. Discard the supernatants.

3.2 Mitochondrial Staining and Flow Cytometry

1. To stain mitochondria, add 100 μL of 150–200 nM MitoTracker Red CMXRos (MTR) and/or MitoTracker Green FM (MTG) previously dissolved in DMSO and diluted in PBS supplemented with FBS. Resuspend cells thoroughly by pipetting up and down and then incubate at 37 °C for 25 min in the dark.
2. Pellet and wash cells as in **step 6**.
3. Add 200 μL of 2–4% paraformaldehyde in PBS and incubate for 15 min at room temperature (*see* **Note 4**).
4. Repeat washing **step 6**. After the last centrifugation step, resuspend cells in FACS buffer and analyze.
5. Excite the cells with appropriate laser and measure emission from 600 to 620 nm (for MTR) or 500–520 nm (for MTG).

4 Notes

1. There are three variations regarding the fluorescent labeling of mitochondria and monitoring of mitophagy flux: one using exclusively MitoTracker dyes; the second using combination of fluorescently labeled LC3 protein to be expressed from the vector and MitoTracker dye; and third, where both fluorescently labeled LC3 and Bnip3 or Bnip3L/Nix are transfected and analyzed. Here, a protocol for MitoTracker Red or Green dyes is described, with the optional transfection step added. If the second or third variation is being performed, one should pay attention that different fluorescent labeling is used (e.g., combination of GFP-LC3 and MitoTracker Red or GFP-LC3 and RFP-Bnip3L/Nix).
2. If exclusively studying Bnip3/Nix-dependent mitophagy and MitoTracker dyes are being used, it is recommended that Bnip3 and/or Bnip3L/Nix endogenous expression be validated by Western blot prior to flow cytometry analysis.
3. Further, to avoid transfection step of the protocol, GFP-LC3 and/or RFP-Bnip3/Nix stable cell lines could be used (GFP-LC3 stable cell lines are commercially available).
4. It is worth mentioning that fixation step of the protocol might be a problem if MitoTracker Green FM is used. To circumvent this, fixation step of the protocol could be excluded. In this case, it is recommended that the cells are analyzed on flow cytometer as soon as possible.

References

1. Ney PA (2015) Mitochondrial autophagy: origins, significance, and role of BNIP3 and NIX. Biochim Biophys Acta 1853:2775–2783
2. Hamacher-Brady A, Brady NR (2015) Bax/Bak-dependent, Drp1-independent targeting of X-linked inhibitor of apoptosis protein (XIAP) into inner mitochondrial compartments counteracts Smac/DIABLO-dependent effector caspase activation. J Biol Chem 290(36):22005–22018
3. Novak I (2012) Mitophagy: a complex mechanism of mitochondrial removal. Antioxid Redox Signal 17(5):794–802
4. Lazarou M et al (2015) The ubiquitin kinase PINK1 recruits autophagy receptors to induce mitophagy. Nature 524(7565):309–314
5. Novak I et al (2010) Nix is a selective autophagy receptor for mitochondrial clearance. EMBO Rep 11(1):45–51
6. Zhu Y et al (2013) Modulation of serines 17 and 24 in the LC3-interacting region of Bnip3 determines pro-survival mitophagy versus apoptosis. J Biol Chem 288(2):1099–1113
7. Murakawa T et al (2015) Bcl-2-like protein 13 is a mammalian Atg32 homologue that mediates mitophagy and mitochondrial fragmentation. Nat Commun 6:7527
8. Wei H, Liu L, Chen Q (2015) Selective removal of mitochondria via mitophagy: distinct pathways for different mitochondrial stresses. Biochim Biophys Acta 1853:2784–2790
9. Mauro-Lizcano M et al (2015) New method to assess mitophagy flux by flow cytometry. Autophagy 11(5):833–843
10. Zhang J, Ney PA (2010) Reticulocyte mitophagy: monitoring mitochondrial clearance in a mammalian model. Autophagy 6(3):405–408
11. Puleston D et al (2015) Techniques for the detection of autophagy in primary mammalian cells. Cold Spring Harb Protoc 2015(9):070391

Methods in Molecular Biology (2018) 1759: 111–121
DOI 10.1007/7651_2017_15

Published online: 24 March 2017

Exploring MicroRNAs on NIX-Dependent Mitophagy

Wen Li, Hao Chen, Shupeng Li, Guanghong Lin, and Du Feng

Abstract

The dysregulation of autophagy is implicated in many pathological disorders including infections, aging, neurodegenerative diseases, and cancer. Autophagy can be precisely controlled both transcriptionally and translationally. Accumulating evidences show that the autophagy response is regulated by microRNAs, which therefore becomes subject area of interest in recent years. Herein, we give a brief introduction of the recent advancement in the regulation of microRNA on autophagy, and then we focus on the microRNA regulation of the mitophagy receptor, NIX. Finally, we present the methodology on how to study it in detail.

Keywords: MicroRNAs, NIX, BNIP3L, Autophagy, Mitophagy, LC3

1 Introduction

Mitochondrion is an important organelle to provide energy for the cell. However, it is also the major source of cellular reactive oxygen species (ROS) that damage cellular components, especially the mitochondria. Mitophagy is a process selectively degradating the excess or damaged mitochondria to maintain cellular homeostasis. Mitophagy contains two classical pathways: one is PINK1-Parkin-mediated mitophagy, and the other is mitophagy receptor-mediated mitophagy.

In yeast, Atg32 is essential for mitophagy. Atg32 interacts with Atg8 and Atg11 to promote core Atg protein assembly for mitophagy [1, 2]. No mammalian homolog has been identified so far. Until recently, FUNDC1 and NIX (also known as BNIP3L) were demonstrated as mitophagy receptors in mammals.

Our previous study demonstrated that FUNDC1 is localized on mitochondria with its amino terminus exposed to the cytosol and carboxy terminus stretched in the intermembrane space. FUNDC1 mediates hypoxia-induced mitophagy through its characteristic LIR to specifically interact with LC3 [3].

NIX, a protein with homology to BCL2 in the BH3 domain, is another mitochondrial outer membrane protein. NIX could induce cell death and autophagy. Until recently, three mechanisms of NIX-dependent mitophagy were identified. The first model of

NIX-dependent mitophagy was considered to trigger mitochondrial depolarization [4, 5]. A second model assumed that NIX could recruit autophagy components to trigger mitochondrial depolarization, independently [6]. The third model of NIX-dependent mitophagy focused on cross talk between autophagy and cell death pathways. BNIP3 and Beclin-1 were identified to interact with BCL2 or BCL-X_L. This interaction triggered completion between NIX and Beclin-1 for binding to BCL-X_L. As known, Beclin-1 regulates autophagy induction [7], and BCL2 inhibits autophagy by competitively binding to Beclin-1 [8]. However, a novel function of NIX, namely, an ability to induce ROS-mediated autophagy and prime Parkin-ubiquitin-P62-mediated mitochondrial, has also been reported [9]. Selective mitochondrial clearance in reticulocytes requires NIX instead of BAX, BAK, BCL-X_L, BIM, or PUMA, indicating a new pathway independent of apoptosis [10].

In the past several years, noncoding RNAs (ncRNAs) have gained a lot of interest as a new way to regulate gene expression. MicroRNAs (miRNAs) are a kind of small (21–23 nt) noncoding RNAs which could regulate gene expression via binding to the 3′-untranslated region (UTR) of a target mRNA, thus blocking mRNA translation or mediating mRNA degradation [11]. At least six continuous complementary nucleotides between 5′-seed of miRNA and 3′ UTR of mRNA are required for the miRNA-mRNA interaction [12]. miRNAs are implicated in development [13, 14], tumor [15], cancer [16, 17], drug resistance [18–20], neurodegenerative diseases [21, 22], allergic reaction [23], lipogenesis [24], and so on. In this chapter, we will introduce: (1.1) selected examples of NIX-dependent mitophagy regulated by microRNAs; (1.2) NIX-dependent mitophagy regulated by microRNAs that play an important role in development; (1.3) step-by-step protocol of microRNA-regulated autophagy.

1.1 Selected Examples of NIX-Dependent Mitophagy Regulated by MicroRNAs

In recent years, miRNAs also emerged as important regulators of mitophagy. Our previous paper suggests that microRNA-137 is a novel hypoxia-responsive microRNA that inhibits mitophagy via regulation of two mitophagy receptors FUNDC1 and NIX. miR-137 impairs the binding of mitophagy receptor to LC3, thereby, inhibiting the targeting of mitochondria to autophagosome [25]. In view of the importance of NIX, we wondered the mechanism of NIX-dependent mitophagy regulated by miRNAs, as well as the role of NIX-dependent mitophagy regulated by microRNAs.

1.2 NIX-Dependent Mitophagy Regulated by MicroRNAs That Play an Important Role in Development

Mounting evidence suggests that miRNAs play an important role in development. miR-133a regulated mitochondrial function through translational inhibition of NIX. A rat model of gestational diabetes was used to define the metabolic alterations in muscle tissue. Rats exposed to diabetes during gestation become insulin resistant. Moreover, exposure to maternal diabetes increased diacylglycerols in the soleus muscle of both low fat (LF) and HFS (high fat and high sucrose)-fed offspring. Consistent with this, miR-133a expression was also reduced. In addition, the expression level of mitochondrial-labeled genes was detected, showing an increased expression of NIX [26].

1.3 NIX-Dependent Mitophagy Regulated by MicroRNAs Involved in Cell Cycle and Apoptosis Pathways

It has been reported that miRNAs play significant roles in regulating human embryonic stem cell (hESC) self-renewal, differentiation, and cell reprogramming [27–29]. Among these miRNAs, miR-302/367 cluster is the most abundant miRNA in hESC. However, the mechanism of miR-302/367 cluster in regulating hESC self-renewal or growth remains largely unknown. MiR-302/367 cluster could regulate cell cycle and apoptosis in hESC. The researchers identified that NIX was the target gene of miR-302/367 cluster, by screening several key cell cycle regulators that were negatively regulated by miR-302/367 cluster [30].

1.4 MicroRNAs Targeted NIX to Regulate Erythropoiesis Through Stage-Specific Control of Mitophagy

Kap1-deleted erythroblasts failed to induce mitophagy-associated genes and retained mitochondria. The KRAB/KAP1-miRNA regulatory cascade is evolutionary conserved, as it also controls mitophagy during human erythropoiesis. MicroRNA expression profile chip of control and Kap1 KO cells indicated that 11 miRNAs were upregulated in KO cells. Further research was taken to discover that six of these upregulated miRNAs might target mitophagy-related genes. Notably, miR-351 was predicted to act on BNIP3L [31].

2 Materials

All cell transfection experiments, including plasmids and RNA, were taken by Lipofectamine 2000 (Invitrogen). The centrifuge model used in all experiments is Eppendorf centrifuge (5424R). All the reagents are stored at room temperature in addition to indicated otherwise.

2.1 Luciferase Assay

We take Dual-Luciferase Assay Kit (Promega) for luciferase assay detection, which is sufficient to perform 100 standard assays. This system contains 10 mL Luciferase Assay Buffer II, 1 vial Luciferase Assay Substrate (lyophilized product), 10 mL Stop & Glo® Buffer, 200 μL Stop & Glo® Substrate (50×), and 30 mL Passive Lysis

Buffer (5×). Working solution should be only prepared 2–3 h before detection to avoid fluorescence quenching.

For 1× PLB: 5× Passive Lysis Buffer is needed to be diluted by 4 volumes distilled water. This 1× PLB should be mixed well and stored at 4 °C.

For LAR II: 1 vial Luciferase Assay Substrate was dissolved by 10 mL Luciferase Assay Buffer II. Packing solution can avoid repeated freezing and thawing.

Stop & Glo® Reagent: 50× mother liquor could be diluted by 10 mL Stop & Glo® Buffer to make a 1× solution of Stop & Glo® Reagent.

2.2 Real-Time PCR

TRIzol (Life Technologies), DNase I, RNase-free (Thermo), SYBR PrimeScript miRNA RT-PCR Kit (TaKaRa), and SYBR Premix Ex Taq™ (Tli RNaseH Plus) (TaKaRa).

2.3 Western Blotting

1. 6× SDS loading buffer (100 mL): 70 mL 4× Tris pH 6.8, 3.8 g glycerol(30%(V/V)), 10 g SDS, 9.3 g DTT, 12 mg bromine phenol blue; store at −20 °C after dissolving.
2. Lysis buffer (500 mL): Adding ultrapure water to dissolve 3.028 g Tris, 4.383 g NaCl, and 0.186 g EDTA, followed by adjusting pH value at 7.5. Later, 5 mL NP40 and 50 mL glycerol were added to make the total solution volume to 500 mL. Store lysis buffer at 4 °C.

3 Methods

Comprehensive and authoritative methods and protocols of miRNA research area will provide researchers, educators, clinicians, and biotechnology specialists with a broad understanding of the issues in miRNA's functional research strategically. This volume provides a collection of the most widely used strategies in miRNA that were developed and refined over the years. They are mainly focused on miRNA target gene prediction and validation of the interaction between miRNA and its target genes, including lists of the necessary materials and reagents, step-by-step, readily reproducible laboratory protocols.

3.1 miRNA's Target Gene Prediction

As described above, miRNAs regulated the expression of a target gene in a sequence-specific manner. In the following section, some of the classic commonly used miRNA target gene prediction software will be presented in Table 1. Hundreds of putative genes may be the target of an individual miRNA, and a single transcript may be regulated by multiple miRNAs.

Traditional microRNA's target gene usually includes these features: (1) base pairing of seed sequence, (2) GC content of

Table 1
Target prediction algorithms

Software	URL link	miRNA → Target gene	Target gene → miRNA
TargetScan	http://www.targetscan.org	Yes	Yes
PicTar	http://pictar.mdc-berlin.de/	Yes	Yes
DIANA LAB	http://83.212.96.7/DianaToolsNew/index.php?r=site/home	Yes	Yes
miRanda	http://www.microrna.org/microrna/home.do	Yes	Yes

seed sequence, (3) conservation of seed sequence, (4) number of hydrogen bonds in the seed zone, (5) minimum free energy, and (6) conservation of target gene sequence. Those features could be used as the screening criteria of the target gene to further filter the results predicted by software mentioned above.

3.2 Validation of the Interaction Between miRNA and its Target Genes

Most computational prediction could have a false positive result due to its essentially scanning the whole transcriptome. To address this problem, some experimental verification must be taken to confirm the interaction between miRNA and the so-called target gene.

Expression profile chips could large-scale screen differentially expressed genes, to obtain the research direction of interested. Big data can give us some guidance on the interaction between miRNA and target genes. However, to confirm the interaction between miRNA and target genes, further experiments are needed to prove it.

3.2.1 Luciferase Assay

Luciferase assay is a classical experimental method to check the reporter activity. Dual reporters are commonly used to improve experimental accuracy. In the Dual-Luciferase® Reporter (DLR™) Assay, the activities of firefly (*Photinus pyralis*) and Renilla (*Renilla reniformis*, also known as sea pansy) luciferases are measured sequentially from a single sample. In the DLR™ Assay System, both reporters yield linear assays with subattomole sensitivities. Furthermore, no endogenous activity of either reporter in the experimental host cells was detected.

Step-by-step laboratory protocols:

1. Construction of luciferase reporter plasmid: 3′UTR relevant fragments of target genes were inserted downstream of the firefly luciferase gene of the pmirGLO vector.
2. Cell preparation and transfection: HeLa and HEK293 cells were commonly used in this part due to their higher transfective efficiency than other cell lines. Cells were seeded in 24-well

plates 1 day before transfection. When the cells were completely spread out and the cell density reached 80%, the cells were transiently transfected with 20 μM scramble NC, miRNA mimics, scramble NC inhibitor, or miRNA inhibitor 2.5 μL, respectively. Eight hours later, 40 ng of WT reporter plasmid or mutant plasmid were also transfected for another 24 h (*see* **Notes 1–3**). Three repeated experiments were designed for each group. Lipofectamine 2000 (Invitrogen) was used for transfection.

3. Cell Lysis: Discard the culture medium and rinse cells in 1× PBS. Remove all rinse solution. Add 100 μL 1× PLB for each well (24-well plates), followed by 15 min lysis on ice. After repeatedly pipetting lysate several times, the lysate was centrifuged at 4 °C for 10,000 rpm, 10 min. After centrifugation, the supernatant was collected for the next step (*see* **Note 4**).
4. Sample detection: Sirius L Tube Luminometer (Berthold Detection Systems) and FB12 Sirius software V 2.0 were used to detect firefly and Renilla luciferase activities. System parameters were set as "delay time: 2 s" and "measurement time: 10 s". 20 μL PLB lysate/well was transferred to each 1.5 mL tube, and 100 μL LAR II was dispensed in the above 1.5 mL tube later. After well mixed, the firefly luciferase activity was detected. Every sample's firefly luciferase activity was recorded. 100 μL Stop & Glo® Reagent was subsequently added in the previously described 1.5 mL tube, to terminate firefly luciferase activity and activate Renilla luciferase activity (*see* **Note 4**). Ensuring the same times of mixing and avoiding the generation of bubbles will reduce the error of system detection. Data used to plot was calculated by the formula: $X =$ firefly luminescent signal/Renilla luminescent signal.

3.2.2 Real-Time PCR

Because the interaction of miRNA with the target gene is reflected in the level of mRNA in some degree, real-time PCR is a commonly used method of verification. Cells were seeded in 6-well plates 1 day before transfection. For real-time PCR, the cells were transiently transfected 20 μM scramble NC, miRNA mimics, scramble NC inhibitor, or miRNA inhibitor 10 μL, respectively, as described above. Total RNA was isolated with TRIzol (Life Technologies), followed by a DNase treatment to eliminate contaminating genomic DNA and reverse transcription reaction. Relative gene expression was determined using two-step quantitative RT-PCR.

Specific experimental steps are as follows:

1. Remove culture medium and rinse cells in PBS (RNase-free).
2. Add 1 mL TRIzol in each well and pipette lysate several times.

3. Cell lysate was transferred to 1.5 mL RNase-free tube, followed by complete vortex and placed at room temperature for 10 min.
4. Add 0.2 mL chloroform and mix thoroughly, followed by placement at room temperature for 15 min.
5. Centrifugation at 4 °C, 12,000 rpm for 10 min.
6. Transfer the upper aqueous phase to a new 1.5 mL RNase-free tube, followed by adding equal volume of isopropanol. Mix thoroughly by vigorously inverting 4–6 times. Incubate at room temperature for 15 min.
7. Centrifugation at 4 °C, 12,000 rpm for 15 min.
8. Discard the supernatant and add 1 mL 75% ethanol (DEPC water diluted). Wash the sediment.
9. Centrifugation at 8,000 rpm at 4 °C for 5 min and discard the supernatant.
10. Repeat 8–9 for one time.
11. Air-dry pellet for 5–10 min, and redissolve RNA in a suitable volume of RNase-free water.
12. For RT-PCR applications, template RNA must be free of DNA contamination. Prior to cDNA synthesis, RNA can be treated with genomic DNA elimination. We apply DNase I, RNase-free (Thermo) to remove trace amounts of DNA. The reaction system and procedure are shown in Table 2.
13. Apply the prepared RNA as a template for reverse transcription reaction. Thermo Scientific RevertAid First Strand cDNA Synthesis Kit was used in this part. The reaction system and procedure are shown in Table 3.
14. Quantitative PCRs were performed with SYBR PrimeScript miRNA RT-PCR Kit (TaKaRa) on a Roche LightCycler480II Real-time PCR System. Amplification and relative quantification of cDNA were carried out with SYBR Premix Ex Taq™ (Tli RNaseH Plus) (TaKaRa) (*see* **Notes 5** and **6**). The reaction system and procedure are shown in Table 4. Fold changes were calculated using the $\Delta\Delta$Ct method with normalization to U6 rRNA endogenous control.

Table 2
Genomic DNA elimination

Reaction component		Procedure
RNA	1 μg	37 °C, 30 min
10× Reaction Buffer with $MgCl_2$	1 μL	Add 1 μL 50 mM EDTA and incubate at 65 °C, 10 min
DNase I, RNase-free	1 μL (1 U)	
Water, nuclease-free	to 10 μL	

Table 3
Reverse transcription reaction

Reaction component		Procedure
Prepared RNA in Table 2	10 μL	42 °C, 60 min
Primer(Oligo dT) or gene-specific primer	1 μL	Terminate the reaction at 70 °C, 5 min
Water, nuclease-free	1 μL	
5× Reaction Buffer	4 μL	
RiboLock Rnase Inhibitor(20 U/pL)	1 μL	
10 mM dNTP Mix	2 μL	
RevertAid M-MuLV RT(200 U/μL)	1 μL	

Table 4
Quantitative PCRs

Reaction component		Procedure	
SYBR Premix Ex Taq(Tli RNaseH Plus) (2×)	4 μL	Stage1: 95 °C, 30 s	Reps: 1
PCR Forward Primer (10 μM)	0.2 μL	Stage2: 95 °C, 5 s → 60 °C, 30 s	Reps: 40
PCR Reverse Primer (10 μM)	0.2 μL	Stage3: Melt Curve	Reps: 1
DNA template	1 μL	95 °C, 5 s → 60 °C, 60 s	
ddH_2O	4.6 μL	Cooling	
Total	10 μL		

3.2.3 Western Blotting

In order to detect the effect of miRNA on target gene at translational level, Western blotting is needed (*see* **Notes 7** and **8**). Whole cell lysates used for immunoblotting were prepared in lysis buffer containing a phosphatase inhibitor (Roche) on ice for 30 min. Lysates were mixed with 6× SDS loading buffer and boiled for 10 min. Protein samples were separated on SDS/PAGE gels and then transferred to PVDF membrane. Membranes were blocked by 5% nonfat milk (dissolved in phosphate-buffered saline with Tween 20, PBST) for 1 h at room temperature. Corresponding primary antibody were used for overnight incubation at 4 °C, followed by HRP-labeled secondary antibodies' incubation at 37 °C for 2 h. For loading controls, membranes were probed with the antibody against β-tubulin or β-actin. Densitometric ratios were quantified by IMAGEJ software (Bethesda, MD, USA).

4 Notes

1. Be careful to handle RNA because it is very easy to be degraded by the RNase everywhere. Do to treat the key solutions with 0.1% DEPC.
2. To avoid the degradation of RNA, microRNA mimics and inhibitors should be preferably packed individually with

sterilized RNase-free PCR tubes after dissolved, to avoid repeated freezing and thawing. If storage solution concentration is 20 μM, then it is probably to be packed 30 μL into one tube and stored at −80 °C.

3. The plasmid should be dissolved in DEPC water (or RNase-free water). When we do the co-transfection of RNA and plasmid, all the tips and EP tubes should be RNase-free.
4. In Luciferase assay, all cell lysis should be placed on ice before sample detection, so as to avoid the degradation of protein. To avoid bubbles in the process of mixing cell lysis, LAR II, and Stop & Glo® Reagent, make sure that samples in each tube were pipetted for the same times. As fast as possible to complete the detection, in order to avoid fluorescence quenching.
5. If you want to use primer repeatedly in short term, the primer can be stored at 4 °C to avoid the degradation by repeated freezing and thawing.
6. Once PCR reaction is inhibited or the efficiency is low, the continuous gradient dilution of template can be considered to evaluate the efficiency of the reaction. As the template gradually diluted, the inhibitory effect disappeared, and the curve also showed a more obvious characteristic exponential shape.
7. In order to improve the transfection efficiency, time for cells to be replaced with full culture solution was prolonged to 12 h after the cells were transfected with exogenous RNA.
8. For protein level detection, the time of exogenous transfection was ensured at least 24 h, or even 36 and 48 h.

Acknowledgement

This work was supported by NSFC (No. 31401182), by the National Basic Research Program of China (2013CB910100), by the Natural Science Foundation of Guangdong Province, China (2014A030313533), by Yangfan Plan of Talents Recruitment Grant, Guangdong, China (Yue Cai Jiao [2015] 216, 4YF14007G), by the Science and Technology Planning Project, Guangdong, China (No. 2016A020215152), by Guangdong Medical Research Foundation (A2015332), and by Research Fund of Guangdong Medical University (M2014024, M2015001).

References

1. Aoki Y, Kanki T, Hirota Y, Kurihara Y, Saigusa T, Uchiumi T, Kang D (2011) Phosphorylation of Serine 114 on Atg32 mediates mitophagy. Mol Biol Cell 22(17):3206–3217
2. Barde I, Rauwel B, Marin-Florez RM, Corsinotti A, Laurenti E, Verp S, Offner S, Marquis J, Kapopoulou A, Vanicek J, Trono D (2013) A KRAB/KAP1-miRNA cascade regulates erythropoiesis through stage-specific control of mitophagy. Science 340(6130):350–353
3. Bartel DP (2009) MicroRNAs: target recognition and regulatory functions. Cell 136 (2):215–233
4. Basak I, Patil KS, Alves G, Larsen JP, Moller SG (2016) microRNAs as neuroregulators, biomarkers and therapeutic agents in neurodegenerative diseases. Cell Mol Life Sci 73 (4):811–827
5. Ding WX, Ni HM, Li M, Liao Y, Chen X, Stolz DB, Dorn GW 2nd, Yin XM (2010) NIX is critical to two distinct phases of mitophagy, reactive oxygen species-mediated autophagy induction and Parkin-ubiquitin-p62-mediated mitochondrial priming. J Biol Chem 285 (36):27879–27890
6. Dwivedi SK, Mustafi SB, Mangala LS, Jiang D, Pradeep S, Rodriguez-Aguayo C, Ling H, Ivan C, Mukherjee P, Calin GA, Lopez-Berestein G, Sood AK, Bhattacharya R (2016) Therapeutic evaluation of microRNA-15a and microRNA-16 in ovarian cancer. Oncotarget 7(12):15093
7. Elmore SP, Qian T, Grissom SF, Lemasters JJ (2001) The mitochondrial permeability transition initiates autophagy in rat hepatocytes. FASEB J 15(12):2286–2287
8. Gomes BC, Rueff J, Rodrigues AS (2016) MicroRNAs and Cancer Drug Resistance. Methods Mol Biol 1395:137–162
9. Gunaratne PH, Creighton CJ, Watson M, Tennakoon JB (2010) Large-scale integration of MicroRNA and gene expression data for identification of enriched microRNA-mRNA associations in biological systems. Methods Mol Biol 667:297–315
10. Li W, Zhang X, Zhuang H, Chen HG, Chen Y, Tian W, Wu W, Li Y, Wang S, Zhang L, Chen Y, Li L, Zhao B, Sui S, Hu Z, Feng D (2014) MicroRNA-137 is a novel hypoxia-responsive microRNA that inhibits mitophagy via regulation of two mitophagy receptors FUNDC1 and NIX. J Biol Chem 289(15):10691–10701
11. Lin SL, Chang DC, Lin CH, Ying SY, Leu D, Wu DT (2011) Regulation of somatic cell reprogramming through inducible mir-302 expression. Nucleic Acids Res 39 (3):1054–1065
12. Liu H (2012) MicroRNAs in breast cancer initiation and progression. Cell Mol Life Sci 69(21):3587–3599
13. Liu L, Feng D, Chen G, Chen M, Zheng Q, Song P, Ma Q, Zhu C, Wang R, Qi W, Huang L, Xue P, Li B, Wang X, Jin H, Wang J, Yang F, Liu P, Zhu Y, Sui S, Chen Q (2012) Mitochondrial outer-membrane protein FUNDC1 mediates hypoxia-induced mitophagy in mammalian cells. Nat Cell Biol 14(2):177–185
14. Miyoshi N, Ishii H, Nagano H, Haraguchi N, Dewi DL, Kano Y, Nishikawa S, Tanemura M, Mimori K, Tanaka F, Saito T, Nishimura J, Takemasa I, Mizushima T, Ikeda M, Yamamoto H, Sekimoto M, Doki Y, Mori M (2011) Reprogramming of mouse and human cells to pluripotency using mature microRNAs. Cell Stem Cell 8(6):633–638
15. Mughal W, Nguyen L, Pustylnik S, da Silva Rosa SC, Piotrowski S, Chapman D, Du M, Alli NS, Grigull J, Halayko AJ, Aliani M, Topham MK, Epand RM, Hatch GM, Pereira TJ, Kereliuk S, McDermott JC, Rampitsch C, Dolinsky VW, Gordon JW (2015) A conserved MADS-box phosphorylation motif regulates differentiation and mitochondrial function in skeletal, cardiac, and smooth muscle cells. Cell Death Dis 6:e1944
16. Okamoto K, Kondo-Okamoto N, Ohsumi Y (2009) Mitochondria-anchored receptor Atg32 mediates degradation of mitochondria via selective autophagy. Dev Cell 17(1):87–97
17. Olde Loohuis NF, Nadif Kasri N, Glennon JC, van Bokhoven H, Hebert SS, Kaplan BB, Martens GJ, Aschrafi A (2016) The schizophrenia risk gene MIR137 acts as a hippocampal gene network node orchestrating the expression of genes relevant to nervous system development and function. Prog Neuropsychopharmacol Biol Psychiatry 73:109
18. Pattingre S, Tassa A, Qu X, Garuti R, Liang XH, Mizushima N, Packer M, Schneider MD, Levine B (2005) Bcl-2 antiapoptotic proteins inhibit Beclin 1-dependent autophagy. Cell 122(6):927–939
19. Rebane A (2015) microRNA and Allergy. Adv Exp Med Biol 888:331–352
20. Sarkar FH, Li Y, Wang Z, Kong D, Ali S (2010) Implication of microRNAs in drug resistance for designing novel cancer therapy. Drug Resist Updat 13(3):57–66
21. Schweers RL, Zhang J, Randall MS, Loyd MR, Li W, Dorsey FC, Kundu M, Opferman JT, Cleveland JL, Miller JL, Ney PA (2007) NIX is required for programmed mitochondrial clearance during reticulocyte maturation. Proc Natl Acad Sci U S A 104(49):19500–19505

22. Tian W, Chen J, He H, Deng Y (2013) MicroRNAs and drug resistance of breast cancer: basic evidence and clinical applications. Clin Transl Oncol 15(5):335–342
23. Twig G, Elorza A, Molina AJ, Mohamed H, Wikstrom JD, Walzer G, Stiles L, Haigh SE, Katz S, Las G, Alroy J, Wu M, Py BF, Yuan J, Deeney JT, Corkey BE, Shirihai OS (2008) Fission and selective fusion govern mitochondrial segregation and elimination by autophagy. EMBO J 27(2):433–446
24. Wystub K, Besser J, Bachmann A, Boettger T, Braun T (2013) miR-1/133a clusters cooperatively specify the cardiomyogenic lineage by adjustment of myocardin levels during embryonic heart development. PLoS Genet 9(9): e1003793
25. Yorimitsu T, Klionsky DJ (2005) Autophagy: molecular machinery for self-eating. Cell Death Differ 12(Suppl 2):1542–1552
26. Zhang J, Ney PA (2009) Role of BNIP3 and NIX in cell death, autophagy, and mitophagy. Cell Death Differ 16(7):939–946
27. Zhang Z, Hong Y, Xiang D, Zhu P, Wu E, Li W, Mosenson J, Wu WS (2015) MicroRNA-302/367 cluster governs hESC self-renewal by dually regulating cell cycle and apoptosis pathways. Stem Cell Rep 4(4):645–657
28. Zhang Z, Xiang D, Heriyanto F, Gao Y, Qian Z, Wu WS (2013) Dissecting the roles of miR-302/367 cluster in cellular reprogramming using TALE-based repressor and TALEN. Stem Cell Rep 1(3):218–225
29. Zhao N, Jin L, Fei G, Zheng Z, Zhong C (2014) Serum microRNA-133b is associated with low ceruloplasmin levels in Parkinson's disease. Parkinsonism Relat Disord 20 (11):1177–1180
30. Zhong D, Huang G, Zhang Y, Zeng Y, Xu Z, Zhao Y, He X, He F (2013) MicroRNA-1 and microRNA-206 suppress LXRalpha-induced lipogenesis in hepatocytes. Cell Signal 25 (6):1429–1437
31. Zou Z, Wu L, Ding H, Wang Y, Zhang Y, Chen X, Chen X, Zhang CY, Zhang Q, Zen K (2012) MicroRNA-30a sensitizes tumor cells to cis-platinum via suppressing beclin 1-mediated autophagy. J Biol Chem 287 (6):4148–4156

Part III

Others

Methods in Molecular Biology (2018) 1759: 125–132
DOI 10.1007/7651_2017_16

Published online: 30 April 2017

Monitoring of Atg5-Independent Mitophagy

Satoko Arakawa, Shinya Honda, Satoru Torii, Masatsune Tsujioka, and Shigeomi Shimizu

Abstract

Mitophagy is a mitochondrial quality control mechanism where damaged and surplus mitochondria are removed by autophagy. There are at least two different mammalian autophagy pathways: the Atg5-dependent conventional pathway and an Atg5-independent alternative pathway; the latter is involved in the erythrocyte mitophagy. In this chapter we describe the various experimental approaches to assess Atg5-indepedndent mitophagy in mammalian cells.

Keywords: Mitophagy, Atg5-independent autophagy, Erythrocyte maturation, Parkin, Selective autophagy

1 Introduction

Autophagy is a catabolic process where cellular contents, including proteins, lipids, and even entire organelles, are digested within lysosomes. Autophagy was previously considered to be a bulk and nonselective process. However, growing lines of evidence indicate the presence of cargo-specific autophagy (selective autophagy) that eliminates specific organelles [1, 2], including mitochondria (mitophagy). Mitophagy is associated with various biological events such as mitochondrial clearance during terminal differentiation of reticulocytes (Figs. 1 and 2).

It is currently accepted that autophagy is driven by more than 30 autophagy-related proteins (Atgs) that are well conserved from yeasts to mammals [3]. In this machinery, two ubiquitin-like conjugation pathways (the Atg5-Atg12 pathway and the microtubule-associated protein 1 light chain 3 (LC3) pathway) are commonly recognized to be important for elongation/closure of isolation membranes and generation of autophagosomes. Conjugation of phosphatidylethanolamine (PE) to LC3 is coupled with translocation of LC3 from cytosol to the isolation membrane, and hence this translocation has been utilized as a reliable marker of autophagy. Despite the crucial role of Atg5 in autophagy, Atg5-deficient mouse embryos develop normally until the perinatal period [4],

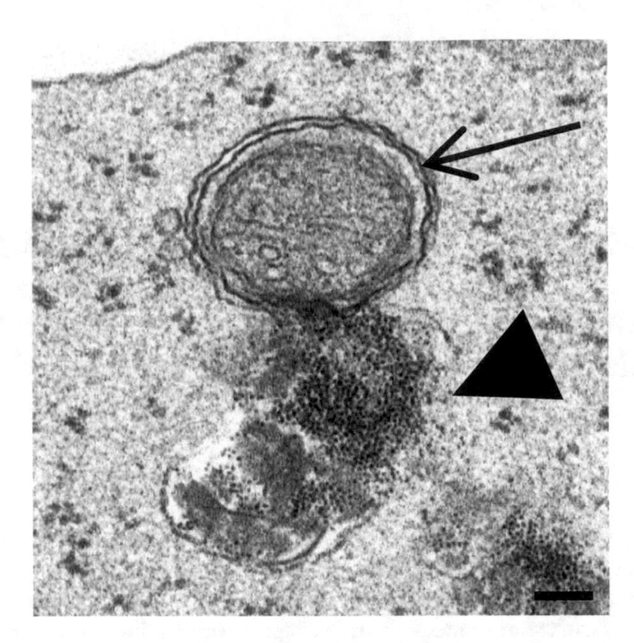

Fig. 1 Clearance of mitochondria by mitophagy from reticulocytes of wild-type (WT) mice. A representative electron micrograph of WT reticulocytes from embryonic liver (E14.5). The *arrow* and *arrowhead* indicate an autophagosome engulfing mitochondrion and an autolysosome degrading mitochondrion, respectively. Note that the autophagosome and autolysosome have double-membrane and single-membrane structures, respectively (bar = 1 μm)

suggesting that an alternative autophagic pathway may exist in such embryos. Actually, we observe typical autophagic structures in Atg5-deficient cells when we add several toxic drugs to the cells [5]. The morphology of Atg5-independent autophagic structures is indistinguishable from that of starvation-induced Atg5-dependent autophagic structures, including the formation of the isolation membrane, the autophagosome, and the autolysosome. The Atg5-independent autophagy requires Unc51-like kinase 1 (Ulk1) and the phosphoinositide 3-kinase complexes, both of which also act at the initiation steps of conventional autophagy. Rab9 is required for the autophagosome membrane closure specifically in the Atg5-independent autophagy.

Until now, Parkinson's disease-associated mitophagy has been well studied in mammals [6]. Among the genes associated with familial Parkinson's disease, Parkin (PARK2) and Pink1 (PARK6) regulate mitophagy for the degradation of damaged mitochondria [7]. This mechanism should be mediated by the Atg5-dependent conventional autophagy, because this mitophagy is largely suppressed by the lack of Atg5 [6]. In contrast, the other representative mitophagy occurring in maturating red blood cells is mediated by the Atg5-independent alternative autophagy [8]. Because this mitophagy occurs in reticulocytes irrespective of the presence of

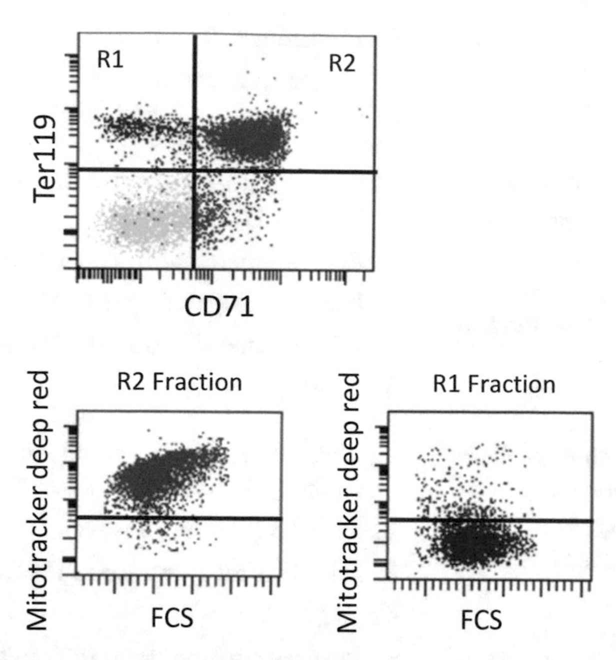

Fig. 2 Detection of mitophagy by three-color flow cytometry. Erythroid cells from WT mice were stained with anti-Ter119, anti-CD71, and MitoTracker Deep Red. (*Upper panel*) Ter119 versus CD71 fluorescence. (*Lower panel*) Forward scatter (FSC) versus MitoTracker fluorescence of the R2 and R1 fractions in *Upper panel*. Almost all the $CD71^{+}Ter119^{+}$ cells (R2 fraction) show the strong MitoTracker staining signal. In contrast, about 90% of the $CD71^{-}Ter119^{+}$ cells (R1 fraction, mature erythrocytes) show the weak MitoTracker staining signal due to mitophagy during differentiation

Atg5, the Atg5-independent autophagy is involved in this type of mitophagy [9].

We here describe the experimental procedures to assess Atg5-indepedndent mitophagy in mouse embryonic fibroblasts (MEFs) and erythrocytes.

2 Materials

2.1 Tandem Tom20-GFP-RFP Fluorescence Assay Using Atg5-Deficient Cells

1. Atg5 (or Atg7) KO MEFs.
2. Tom20-GFP-RFP mammalian expression plasmid.
3. Neon or Amaxa electroporation systems and their equipments.
4. Fluorescence microscope.

2.2 Transmission Electron Microscopy (TEM) Using Atg5-Deficient MEFs

1. Atg5 (or Atg7) KO MEFs.
2. Karnovsky fixative: 1.6% paraformaldehyde and 3% glutaraldehyde in 0.1 M phosphate buffer pH 7.4 with 0.5 mM EGTA (made just before use) (*see* ref. 10).

3. 1% osmium/H_2O.
4. 1% uranyl acetate.
5. Epoxy resin.

2.3 Flow Cytometric Analysis of Atg5-Independent Mitophagy During Erythrocyte Maturation

1. Staining buffer: PBS containing 3% fetal bovine serum and 2 mM EDTA.
2. PE-conjugated anti-Ter119 antibody (BD bioscience).
3. FITC-conjugated anti-CD71 antibody (BD bioscience).
4. MitoTracker Deep Red (Thermo Fisher).
5. Flow cytometer.

2.4 Analysis of Atg5-Independent Mitophagy in Erythrocytes Using Cell Sorter

1. Magnet beads conjugated with anti-Ter119 antibody (BD bioscience, BD Imag cell separation system).
2. Syto-16 (Thermo Fisher).
3. MitoTracker Deep Red (Thermo Fisher).
4. Cell sorter.
5. Primary antibodies: anti-Tom20 (Santa Cruz) and anti-VDAC (Cell Signaling Technology).

3 Methods

3.1 Tandem Tom20-GFP-RFP Fluorescence Assay Using Atg5-Deficient Cells

Tom20 is a mitochondrial outer membrane protein, and hence this fusion protein localizes on mitochondria. The GFP signal is quenched in the acidic compartments (autolysosomes), whereas RFP signal is stable. Therefore, the presence of RFP signal without GFP (red signal) indicates the induction of mitophagy:

1. The Tom20-GFP-RFP plasmid is transfected into Atg5-deficient MEFs (*see* **Note 1**) using the Neon electroporation system according to the supplier's protocol. Amaxa electroporation system is also available (*see* **Note 2**). In order to assess the Atg5-independent mitophagy, it is essential to use Atg5 (or Atg7)-deficient MEFs.
2. Transfected cells are plated on a glass bottom dish.
3. The cells are treated with various chemicals. Then, they are observed using fluorescence microscopy (*see* **Note 3**).
4. Atg5-independent mitophagy can be assessed by the extent of red signals.

3.2 TEM Using Atg5-Deficient MEFs

TEM is the best method to detect the presence of mitophagy. Mitophagy can be detected by the presence of double-membrane autophagosomes enclosing mitochondria and single-membrane autolysosomes digesting mitochondria in Atg5-deficient MEFs:

1. Atg5-deficient MEFs are grown on glass coverslips (Φ12–15 mm) in 12-well dishes (*see* **Note 4**).
2. Aspirate the medium and immerse the glass coverslips in modified Karnovsky fixative for 20 min at room temperature, and rinse them three times with distilled water (*see* **Note 5**).
3. Postfix the cells on the coverslips by 1% osmium/H_2O for 15 min at 4 °C (on ice), and rinse them three times with distilled water.
4. For en bloc staining, place the coverslips in 1% uranyl acetate for 2 h at RT, and rinse them three times with distilled water.
5. Dehydrate the samples in 50% ethanol.
6. Dehydrate the samples completely in 100% ethanol three times (over 30 min in total).
7. To infiltrate an epoxy resin into the samples, immerse the coverslips carrying the fixed cells in the following solutions in a glass petri dish sequentially: a mixed solution of propylene oxide and ethanol (1:1) for 5 min, a mixed solution of propylene oxide and an epoxy resin (1:1) for 5 min, and an epoxy resin for 5 min.
8. Fill the embedding molds with a fresh epoxy resin.
9. Put the coverslip carrying the samples on the microscope slide, and cover the samples with the epoxy resin by quickly inverting the mold on the coverslip.
10. Polymerize the epoxy resin in an oven at 60 °C for 11–12 h. Then, soak the mold attaching the coverslip in liquid nitrogen to transfer the samples from the coverslip to the epoxy resin in the mold (*see* **Note 6**).
11. Polymerize the epoxy resin in the mold again in an oven at 60–70 °C for over 12 h.
12. Observe samples using TEM.

3.3 Flow Cytometric Analysis of Atg5-Independent Mitophagy During Erythrocyte Maturation

The mitophagy during erythrocyte maturation occurs mainly in an Atg5-independent manner. Therefore, Atg5-deficient mice are not required. This mitophagy can be assessed by three-color flow cytometry using a marker of erythroid precursors (CD71), a marker of late-stage erythroid lineage (Ter119), and mitochondria-specific MitoTracker Deep Red staining. In the normal development, almost all the $CD71^{+}Ter119^{-}$ cells and $CD71^{+}Ter119^{+}$ cells possess mitochondria and thereby show the strong MitoTracker staining signal. In contrast, most $CD71^{-}Ter119^{+}$ cells (mature erythrocytes) show the weak MitoTracker staining signal due to the induction of mitophagy:

1. Sacrifice mice and remove skin and muscle around femurs and tibias.

2. Cut off the hip joint, the knee joint, and the end of tibias for the isolation of bone marrow cells.
3. Flush out the bone marrow cells into the 1.5 ml tube until the color of the bones turn from red to white (*see* **Note 7**).
4. Collect the cells by centrifugation at 700 × *g* for 5 min at 4 °C, resuspend them in PBS, and filter the cell suspension through a 70 μm cell strainer.
5. Mature red blood cells are collected from the caudal vena cava using heparinized syringe and heparinized tubes.
6. The collected cells are suspended with the staining buffer, centrifuged at 700 × *g* for 5 min at 4 °C, and resuspended with the staining buffer.
7. 1×10^6 cells are incubated with PE-conjugated anti-Ter119 antibody (1:200), FITC-conjugated anti-CD71 antibody (1:200), and MitoTracker Deep Red (100 nM) for 30 min at 4 °C.
8. Wash the cells with the staining buffer and analyze using flow cytometer.
9. Mitochondrial clearance is measured by the fraction of MitoTracker Deep Red^{low} cells in CD71^{low}, Ter119^{high} mature erythrocytes.

3.4 Analysis of Atg5-Independent Mitophagy in Erythrocytes Using Cell Sorter

1. Bone marrow cells are harvested as described above.
2. Ter119-positive erythroid cells are isolated using magnet beads conjugated with anti-Ter119 antibody, according to the manufacturer's instructions.
3. Incubated the cells with Syto-16 (50 nM) and MitoTracker Deep Red (100 nM) for 30 min at 4 °C.
4. Isolate erythroblasts (Syto16^{high} MitoTracker^{high}) and reticulocytes (Syto16^{low} MitoTracker^{high}) using a cell sorter.
5. Circulating mature red blood cells are also obtained from the caudal vena cava and isolated using magnet beads conjugated with anti-Ter119 antibody.
6. The extent of mitophagy in the sorted cells is analyzed by the immunoblot analysis as follows.
7. Homogenize cells in a SDS sample buffer.
8. Boil the homogenized samples for 5 min, and electrophorese them using SDS-polyacrylamide gels (*see* **Note 8**).
9. The gels are transferred to PVDF membranes. The membranes are then blocked with 3% skim milk in TBS buffer containing 0.1% Tween-20 (TBS-T) and are reacted with the primary antibody (e.g., anti-Tom20 and anti-VDAC antibodies) overnight at 4 °C.

10. After washing in TBS-T, the membranes are reacted with the horseradish peroxidase-labeled secondary antibody and visualized.

4 Notes

1. MEFs are generated from mouse embryos at embryonic day 13.5 and are immortalized with SV40 T antigen. MEFs are cultured in Dulbecco's Modified Eagle's Medium (DMEM) supplemented with 2 mM of L-glutamine, 1 mM of sodium pyruvate, 0.1 mM of nonessential amino acids, 10 mM of HEPES/Na^+ (pH 7.4), 0.05 mM of 2-mercaptoethanol, 100 U/ml of penicillin, 100 μg/ml of streptomycin, and 10% of fetal bovine serum.
2. For transfection, set Neon and Amaxa programs to voltage 1300, width 20, number 2, and U-20 (V-kit).
3. Tom20-GFP-RFP is normally detected as yellow signals on mitochondria resulting from the merging of green and red signals. When mitophagy occurs, green signal is quenched by lysosomal acidification, and only red fluorescence is detected.
4. Glass bottom dishes are not suitable because the plastic and the paste sticking the glass dissolve in propylene oxide (used in Section 3.2, **step** 7).
5. To make 25 ml modified Karnovsky (1.6% paraformaldehyde, 3% glutaraldehyde in 0.1 M phosphate buffer (pH 7.4) with EGTA 0.5 mM), add 0.4 g paraformaldehyde powder and 0.60 ml NaOH solution (0.1 N) to 2.50 ml distilled water in a 50 ml beaker. Then, warm the mixture by floating the 50 ml beaker in a 200 ml beaker containing hot water (~70 °C). Agitate the mixture gently for approximately 10 min until it becomes clear. Remove the 50 ml beaker from hot water, and add 12.5 ml of 0.2 M phosphate buffer (pH 7.4) containing 1 mM EGTA to the solution. After the solution has returned to room temperature, add 9.4 ml of 8% glutaraldehyde (use EM glade in a 10-ml ampule) to the solution.
6. Immediately after taking samples from an oven, soak samples into liquid nitrogen in a Styrofoam box. Pull the embedding molds from the microscope slide using tweezers.
7. We use the 27-gauge needle attached to the 1 ml syringe filled with the staining buffer.
8. 15% SDS-polyacrylamide gels are recommended.

References

1. Komatsu M, Ichimura Y (2010) Selective autophagy regulates various cellular functions. Genes Cells 15:923–933
2. Ding WX, Yin XM (2012) Mitophagy: mechanisms, pathophysiological roles, and analysis. Biol Chem 393:547–564
3. Mizushima N, Yoshimori T, Ohsumi Y (2011) The role of Atg proteins in autophagosome formation. Annu Rev Cell Dev Biol 27:107–132
4. Kuma A, Hatano M, Matsui M et al (2004) The role of autophagy during the early neonatal starvation period. Nature 432:1032–1036
5. Nishida Y, Arakawa S, Fujitani K et al (2009) Discovery of Atg5/Atg7-independent alternative macroautophagy. Nature 461:654–658
6. Narendra D, Tanaka A, Suen DF et al (2008) Parkin is recruited selectively to impaired mitochondria and promotes their autophagy. J Cell Biol 183:795–803
7. Jin SM, Lazarou M, Wang C et al (2010) Mitochondrial membrane potential regulates PINK1 import and proteolytic destabilization by PARL. J Cell Biol 191:933–942
8. Kent G, Minick OT, Volini FI et al (1966) Autophagic vacuoles in human red cells. Am J Pathol 48:831–857
9. Honda S, Arakawa S, Nishida Y et al (2014) Ulk1-mediated Atg5-independent macroautophagy mediates elimination of mitochondria from embryonic reticulocytes. Nat Commun 5:4004
10. Karnovsky MJ (1965) A formaldehyde-glutaraldehyde fixative of high osmolality for use in electron microscopy. J Cell Biol 27:137A

Methods in Molecular Biology (2018) 1759: 133–140
DOI 10.1007/7651_2017_17

Published online: 20 April 2017

Monitoring of Paternal Mitochondrial Degradation in *Caenorhabditis elegans*

Miyuki Sato and Ken Sato

Abstract

In *Caenorhabditis elegans* embryos, paternally inherited mitochondria and their mitochondrial DNA are degraded via selective autophagy called allophagy (allogeneic organelle autophagy). This is a developmentally programmed autophagy and combined with *C. elegans* genetics and in vivo imaging provides a unique opportunity to analyze selective autophagy under physiological conditions.

Keywords: *C. elegans*, Allophagy, Autophagy, Mitochondria, Mitochondrial DNA, Maternal inheritance

1 Introduction

Maternal (uniparental) inheritance of mitochondrial DNA (mtDNA) is generally observed in many organisms, including mammals. In *C. elegans*, sperm-derived paternal mitochondria and their mtDNA enter the oocytes, but are selectively degraded via autophagy. In addition to paternal mitochondria, membranous organelles (MOs), which are sperm-specific post-Golgi organelles, are simultaneously targeted for autophagic degradation. This selective degradation of paternal organelles was named as allogeneic (non-self) organelle autophagy or allophagy (Fig. 1) [1–4]. *C. elegans* has two sexes, hermaphrodite and male. Hermaphrodite has both oocytes and sperm and produces self-fertilized progeny. When hermaphrodites are outcrossed with males, oocytes from hermaphrodites preferentially fertilize with sperm from males, producing cross-fertilized progeny. Allophagy is observed in both self- and cross-fertilized embryos. In addition to allophagy, it was reported that proteasome is also somehow involved in the elimination of paternal mtDNA [5].

Atg8/LC3 family proteins are generally utilized as markers of autophagosome membranes. *C. elegans* has two homologs of these proteins, LGG-1 and LGG-2, which are sequentially recruited to autophagosomes [6, 7]. Their green fluorescent protein (GFP) fusions expressed in oocytes or antibodies were used to observe

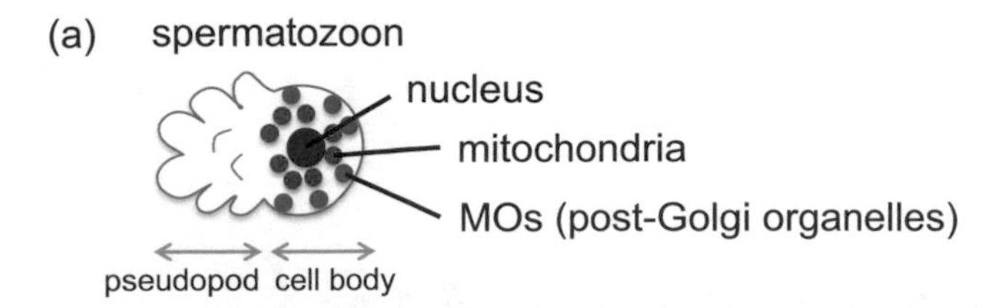

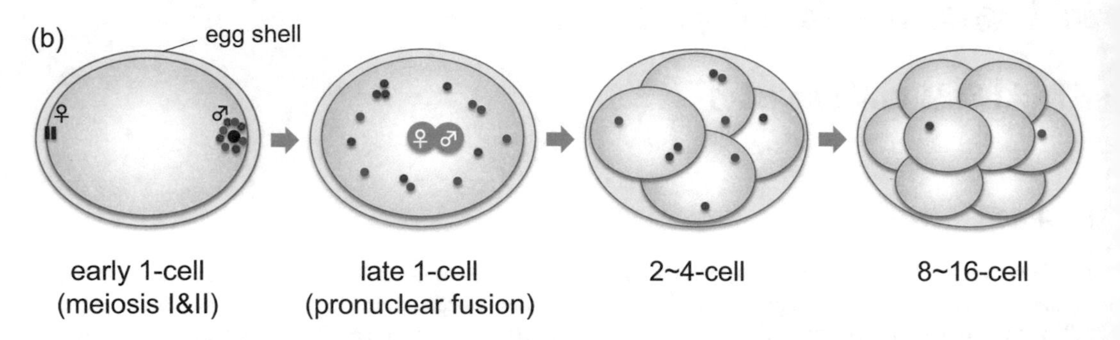

Fig. 1 Schematic diagram of a *C. elegans* spermatozoon and early embryos. (**a**) A *C. elegans* spermatozoon is constituted of a cell body and an extended pseudopod. In the cell body, it contains mitochondria and membranous organelles (MOs) (50–70 per spermatozoon). (**b**) The typical appearance of paternal mitochondria and MOs in the wild-type embryos. In early 1-cell stage embryos, paternal organelles cluster around the penetrating sperm pronucleus. At the meiosis II stage, autophagosomes are formed around these paternal organelles. During the pronuclear expansion and fusion stages, they disperse into the cytoplasm by cytoplasmic streaming. The signals of paternal organelles gradually decrease, and most signals disappear by the 8–16-cell stage. In the 16-cell stage embryos, a small number of paternal organelles are only occasionally observed

autophagosomes in embryos [1, 2]. Paternal mitochondria can be visualized by using strains expressing fluorescent mitochondrial markers, specifically in the sperm. Alternatively, males that are stained with MitoTracker Red are crossed with unstained hermaphrodites [1, 2]. Using either method, paternal mitochondria with fluorescent signals are observable in spermatozoa and F1 fertilized embryos. *C. elegans* spermatozoa have 50–70 mitochondria with unique granular morphology (Fig. 1a) [8]. In early 1-cell stage embryos, during meiosis I and II of maternal DNA, paternal mitochondria and MOs cluster around the condensed paternal pronucleus (~30 min after fertilization). During meiosis II, autophagosomes are formed around these paternal organelles. In the following pronuclear expansion and fusion stages, paternal organelles enclosed by autophagosomes disperse in the cytoplasm via bulk cytoplasmic streaming (30–60 min after fertilization). Paternal organelles are randomly distributed to blastomeres, and they gradually disappear by the 8–16-cell stage (Fig. 1b). In autophagy mutants, such as *lgg-1* or *unc-51*, degradation of paternal mitochondria is inhibited, and the remaining paternal mitochondria are observed beyond the 16-cell stage embryos and even in L1 larvae.

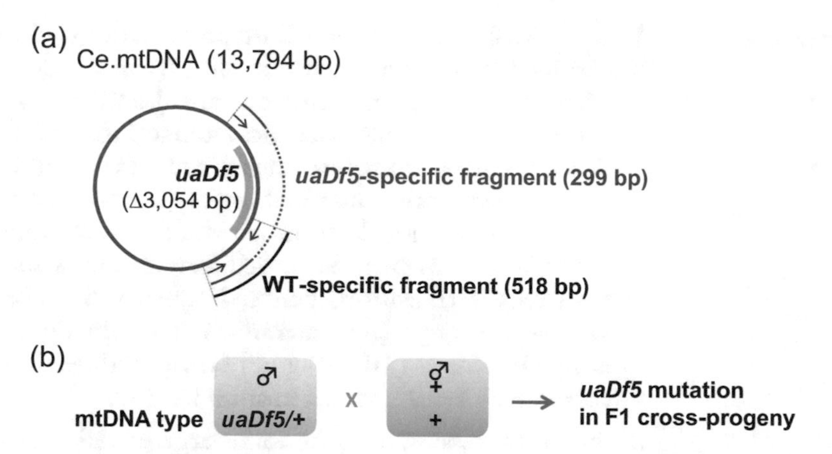

Fig. 2 Detection of paternal mtDNA in the F1 cross-progeny. (**a**) *uaDf5* is a large deletion allele of *C. elegans* mtDNA. Using three primers (*red arrows*), the wild-type and *uaDf5*-specific fragments are amplified by PCR. (**b**) Scheme of mating. Males with *uaDf5/+* and hermaphrodites with the wild-type mtDNA are crossed, and the transmission of *uaDf5* to the F1 generation is examined

Degradation of paternal mitochondria can also be monitored by detecting paternal mtDNA transmission. *uaDf5* is a large deletion allele of mtDNA and is stably kept in a heteroplasmic state with the wild-type mtDNA (in each individual about 60% of mtDNA harbors the deletion) [9]. The mutant and wild-type mtDNA are thought to be together in the same mitochondrion. This *uaDf5* allele is utilized to mark male mtDNA [1, 2]. By detecting male-derived *uaDf5* in the F1 cross-progeny, transmission of paternal mtDNA is monitored (Fig. 2).

2 Materials

2.1 Maintenance of Worm Strains

The general methods used for the handling and culturing of *C. elegans* have been described previously [10, 11]. *C. elegans* is cultured on NGM agar plates seeded with *E. coli* OP50 at 20 °C. Some transgenic strains are incubated at 25 °C, because the expression of some transgenes driven by the oocyte-specific promoter is enhanced at 25 °C.

1. NGM agar plate: Mix 3 g NaCl, 17 g agar, and 2.5 g peptone in 975 mL H_2O. After autoclaving, allow to cool down to ~55 °C, and add 1 mL of 1 M $CaCl_2$, 1 mL of 1 M $MgSO_4$, 1 mL of 5 mg/mL cholesterol in ethanol, and 25 mL of 1 M KPO_4 buffer (pH 6.0). Dispense into 6-cm dishes.
2. *E. coli* OP50 is used as a food source. Culture OP50 in 2xYT medium. Apply about 200 μL of OP50 culture to NGM plates to create a lawn ~3 cm in diameter. For mating, create a smaller lawn (~1 cm in diameter) using ~30 μL culture, because *C. elegans* tends to stay on the *E. coli* lawn; therefore, a smaller lawn increases mating efficiency.

2.2 Worm Strains

1. N2 Bristol is used as the wild-type reference strain. N2 can be obtained from the Caenorhabditis Genetic Center (CGC). Many other mutant strains can also be obtained from CGC and the Japanese National Bioresource Project for nematode. When hermaphrodites are maintained by self-fertilization, they produce exclusively hermaphrodite progeny with rare males (~0.2%). Cross-fertilization produces equal populations of males and hermaphrodites. Self-fertilization is more efficient than cross-fertilization; hence, frequency of males gradually decreases through the generations. To maintain males, manually picked males (10–20) and hermaphrodites (~5) are placed on the NGM plate, which ensures mating.
2. Strains expressing mitochondria-targeted GFP in sperm.

   pwIs623[Pspe-11::hsp-6::GFP, unc-119(+)] [1] and VIG03 *[ant-1.1::GFP, unc-119(+)]* [2] were used. These strains were obtained by the microparticle bombardment method, and *unc-119(+)* was co-introduced as a marker of transformation [12].
3. LB138M; *him-8(e1489); uaDf5/+*
   This strain harbors the *uaDf5* and wild-type heteroplasmic mtDNA [9]. *uaDf5* is stably inherited in self-progeny of hermaphrodites. The *him-8(e1489)* mutation increases male frequency. This strain is deposited in CGC.

2.3 Fluorescent Microscopy

1. M9 buffer: 6 g Na_2HPO_4, 3 g KH_2PO_4, 5 g NaCl, and 0.25 g $MgSO_4 \cdot 7H_2O$ in 1 L H_2O. Sterilize by autoclaving.
2. 2% agarose (electrophoresis grade) in M9 buffer. Using this solution, make thin agar pads on slide glasses (about 1.5 cm in diameter). Living worms and embryos are mounted on these agar pads.
3. 10–20 mM levamisole in M9 buffer. To observe living worms under microscopes, worms are anesthetized by this solution.
4. MitoTracker Red CMXRos, a fluorescent dye that specifically stains mitochondria depending on the membrane potential (Thermo Fisher Scientific, Waltham, MA, USA).
5. Depressed slide glasses, which are used to stain males with MitoTracker.
6. Glass micropipettes, such as Microcaps (Drummond Scientific Company, Broomall, PA, USA).

2.4 Detection of uaDf5 by PCR

1. Wild-type mtDNA and *uaDf5* are simultaneously detected by PCR using three primers:
   U1-F, 5′-CCATCCGTGCTAGAAGACAA-3′

   Cemt1A-R, 5′-CTTCTACAGTGCATTGACCTAGTC-3′

   Cemt5012-F, 5′-TTGGTGTTACAGGGGCAACA-3′

Cemt5012-F (inside the deletion) and Cemt1A-R amplify only the wild-type mtDNA fragment (518 bp). Short extension time is used so that the primers outside the deletion (U1-F and Cemt1A-R) amplify deleted *uaDf5* allele (299 bp) but not the wild-type fragment (3353 bp; Fig. 2) [1, 13].

2. Worm lysis buffer: 10 mM Tris–HCl (pH 8.3), 50 mM KCl, 2.5 mM $MgCl_2$, 0.45% NP-40, 0.45% Tween-20, and 0.01% gelatin. Add Proteinase K (200 μg/mL to the final concentration) before use.
3. Reagents for regular PCR. We use DreamTaq (Thermo Fisher Scientific).

3 Methods

3.1 Monitoring of Paternal Mitochondria Degradation in Self-Fertilized Embryos

By using hermaphrodites expressing mitochondrial markers in sperm, paternal mitochondria in early embryos are easily visualized under a fluorescent microscope or a confocal laser microscope. Early embryos (typically up to 100-cell stage embryos) are kept in the adult hermaphrodite uterus (Fig. 3). They can be observed in intact worms mounted on the agar pad, but it is easier to observe extruded embryos (*see* **Notes 1** and **2**).

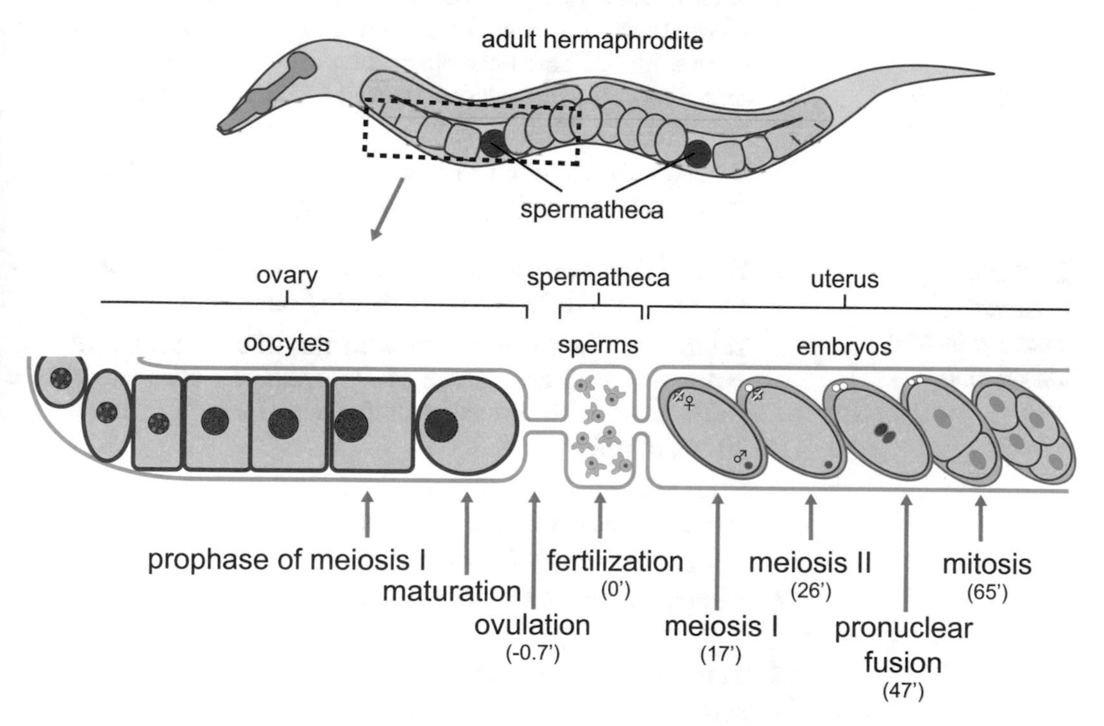

Fig. 3 Schematic diagram of *C. elegans* gonads. *C. elegans* hermaphrodites have two U-shaped gonads. Growing oocytes are aligned in the ovary and the most proximal oocyte is pushed into the spermatheca where it is fertilized. The fertilized oocyte is moved to the uterus and starts developing. The landmark events and their typical time courses are indicated [14]

1. Young gravid adults (10–20) are suspended in a drop of 10–20 mM levamisole in M9 buffer on the coverslip.
2. Under a dissecting stereomicroscope, the middle of the adult body is cut by crossing two 27-G needles. Embryos are extruded.
3. An agar pad on a slide glass is placed on the coverslip softly to sandwich embryos between the coverslip and the agar pad.
4. Embryos are observed under a fluorescent microscope or a confocal laser microscope (*see* **Note 3**).

3.2 Monitoring of Paternal Mitochondria Stained with MitoTracker Red (MT) in Cross-Fertilized Embryos

1. Young adult males are manually collected and transferred to 150 μL M9 buffer containing 2 μM MT on depressed slide glasses. They are incubated at 20 °C for 4 h in a dark and humid chamber.
2. They are transferred to the regular NGM plate using glass micropipettes, such as Microcaps. Do not use plastic tips because worms are sticky. Plate is incubated at 20 °C for 1 h in the dark.
3. Approximately 25 unstained young adult hermaphrodites and excess MT-labeled males (70–100) are placed on a NGM plate with a small OP50 lawn. Plate is incubated at 20 °C, overnight.
4. Mated hermaphrodites accumulate male-derived sperm in the spermatheca, resulting in the red fluorescent signal from the spermatheca. These hermaphrodites are selected under fluorescent dissecting microscope, such as MZ16 FA (Leica Microsystems Japan, Tokyo, Japan).
5. Embryos are extruded from the hermaphrodites and observed as described above (*see* **Notes 4** and **5**).

3.3 Monitoring of Paternal Mitochondrial DNA Transmission

1. Young adult males are collected from population of *uaDf5/+* and stained with MT as described in Section 3.2.
2. Young adult hermaphrodites with the wild-type mtDNA and excess males harboring the *uaDf5* mutation are placed on a NGM plate with a small OP50 lawn. Plate is incubated at 20 °C, overnight.
3. Crossed hermaphrodites containing MT-stained sperm in their spermatheca are transferred to new NGM plates and allowed to lay eggs (*see* **Notes 4** and **5**).
4. F1 eggs or hatched larvae are collected in 2.5 μL lysis buffer in PCR tubes.
5. Tubes are frozen at −80 °C at least for 15 min. Samples can be kept at −80 °C.
6. To lyse embryos or larvae, tubes are incubated at 60 °C for 1.5 h.

7. Proteinase K is inactivated by incubating tubes at 95 °C for 15 min and then kept at 4 °C.
8. PCR solution containing three primers, dNTPs, polymerase, and appropriate buffer is prepared and added directly into the PCR tubes to perform PCR. We usually perform PCR on a total volume of 25 μL, using the DreamTaq and the following conditions:

   95°C , 1 min → (95°C, 30 s → 52°C, 30 s → 72° C, 40 s) × 35 cycles
9. Bands are detected in agarose gel electrophoresis (*see* **Note 6**).

4 Notes

1. Very early 1-cell stage embryos at meiosis I are sensitive to osmotic stress and stop developing when extruded. Hence, they should be observed immediately after dissection or after fixation. 1-cell stage embryos beyond this stage can proceed to normal development after dissection.
2. For fixation, dissected embryos are stuck to MAS-coated slide glasses (Matsunami glass IND., LTD.) and fixed in methanol at −20 °C for 5 min.
3. To quantify remaining paternal mitochondria, z-stacks of confocal images covering the embryos (1 μm intervals) are obtained, and then projected images of confocal z-stacks are generated. These images are transferred to ImageJ or Metamorph software, and the total area giving fluorescent signals per embryo is quantified. In the early 1-cell stage embryos, paternal organelles cluster close together, and it is difficult to quantify each structure. Therefore, for quantification in the 1-cell stage embryos, we use the pronuclear expansion and fusion stage embryos, at which stage paternal organelles scatter in the cytoplasm.
4. When paternal mitochondria are monitored in cross-progeny, the genotype of male also contributes to the results. For example, when *lgg-1(KO)* hermaphrodites and *lgg-1(KO)* males are crossed, paternal mitochondria remain even in the L1 larvae. When *lgg-1(KO)* hermaphrodites and wild-type males are crossed, allophagy at the 1-cell stage is inhibited, and paternal mitochondria remain until mid-stage embryos (around 64-cell stage). However, they disappear in late-stage embryos, because the paternal wild-type *lgg-1* locus expresses zygotically and rescues the autophagy defect.
5. When paternal mitochondria and their mtDNA are monitored in cross-progeny, hermaphrodites crossed with males are

visually selected to eliminate self-progeny of unmated hermaphrodites. After mating, male sperm dominate hermaphrodite sperm. Even when mated hermaphrodites are selected, embryos laid in the beginning are avoided because they might be self-fertilized embryos produced before mating.

6. In the *lgg-1(KO)* background, sperm-derived *uaDf5* in single L1 larva is detectable by PCR.

Acknowledgments

This research was supported by MEXT KAKENHI (Grant Number 26111503), the Cell Science Research Foundation, and the Uehara Memorial Foundation (to M.S.) and by the JSPS KAKENHI (Grant Number 26291036), Sumitomo Foundation, Naito Foundation, the Uehara Memorial Foundation, and Mochida Memorial Foundation for Medical and Pharmaceutical Research (to K.S.).

References

1. Sato M, Sato K (2011) Degradation of paternal mitochondria by fertilization-triggered autophagy in *C. elegans* embryos. Science 334:1141–1144
2. Al Rawi S, Louvet-Vallee S, Djeddi A, Sachse M, Culetto E, Hajjar C, Boyd L, Legouis R, Galy V (2011) Postfertilization autophagy of sperm organelles prevents paternal mitochondrial DNA transmission. Science 334:1144–1147
3. Sato M, Sato K (2012) Maternal inheritance of mitochondrial DNA: degradation of paternal mitochondria by allogeneic organelle autophagy, allophagy. Autophagy 8:424–425
4. Al Rawi S, Louvet-Vallee S, Djeddi A, Sachse M, Culetto E, Hajjar C, Boyd L, Legouis R, Galy V (2012) Allophagy: a macroautophagic process degrading spermatozoid-inherited organelles. Autophagy 8:421–423
5. Zhou Q, Li H, Xue D (2011) Elimination of paternal mitochondria through the lysosomal degradation pathway in *C. elegans*. Cell Res 21:1662–1669
6. Manil-Segalen M, Lefebvre C, Jenzer C, Trichet M, Boulogne C, Satiat-Jeunemaitre B, Legouis R (2014) The *C. elegans* LC3 acts downstream of GABARAP to degrade autophagosomes by interacting with the HOPS subunit VPS39. Dev Cell 28:43–55
7. Djeddi A, Al Rawi S, Deuve JL, Perrois C, Liu YY, Russeau M, Sachse M, Galy V (2015) Sperm-inherited organelle clearance in *C. elegans* relies on LC3-dependent autophagosome targeting to the pericentrosomal area. Development 142:1705–1716
8. Ward S, Argon Y, Nelson GA (1981) Sperm morphogenesis in wild-type and fertilization-defective mutants of *Caenorhabditis elegans*. J Cell Biol 91:26–44
9. Tsang WY, Lemire BD (2002) Stable heteroplasmy but differential inheritance of a large mitochondrial DNA deletion in nematodes. Biochem Cell Biol 80:645–654
10. Brenner S (1974) The genetics of *Caenorhabditis elegans*. Genetics 77:71–94
11. Epstein HF, Shakes DC (eds) (1995) *Caenorhabditis elegans*: modern biological analysis of an organisms, Methods in cell biology, vol 48. Academic Press, San Diego, CA
12. Praitis V, Casey E, Collar D, Austin J (2001) Creation of low-copy integrated transgenic lines in *Caenorhabditis elegans*. Genetics 157:1217–1226
13. Liau WS, Gonzalez-Serricchio AS, Deshommes C, Chin KL, Munyon CW (2007) A persistent mitochondrial deletion reduces fitness and sperm performance in heteroplasmic populations of *C. elegans*. BMC Genet 8:8
14. McCarter J, Bartlett B, Dang T, Schedl T (1999) On the control of oocyte meiotic maturation and ovulation in *Caenorhabditis elegans*. Dev Biol 205:111–128

Methods in Molecular Biology (2018) 1759: 141–149
DOI 10.1007/7651_2017_19

Published online: 22 March 2017

Detection of Hypoxia-Induced and Iron Depletion-Induced Mitophagy in Mammalian Cells

Shun-ichi Yamashita and Tomotake Kanki

Abstract

Mitochondrial quality and quantity are not only regulated by mitochondrial fusion and fission but also by mitochondria degradation. Mitophagy, an autophagy specific for damaged or unnecessary mitochondria, is believed to be an important pathway for mitochondrial homeostasis. To date, several stimuli are known to induce mitophagy. Some of these stimuli, however, including hypoxia, iron depletion, and nitrogen starvation, induce mild mitophagy, which is difficult to detect through decreased mitochondrial mass. Recently, we have clearly detected mitophagy induced under these conditions using mito-Keima as a reporter. In this chapter, we describe the protocols for induction and detection of hypoxia-induced and iron depletion-induced mitophagy using mito-Keima-expressed cells.

Keywords: Mitochondria, Autophagy, Mitophagy, Hypoxia, Iron depletion, Deferiprone, Keima

1 Introduction

Mitochondria are essential organelles that function in important cellular events and numerous metabolic processes, including ATP production, calcium buffering, apoptosis regulation, and fatty-acid β-oxidation. Cellular ATP is primarily produced by oxidative phosphorylation complexes of the mitochondrial inner membrane. In this process, electrons are transported to oxygens by the electron transport chain and form the discontinuous proton concentration across the mitochondrial inner membrane, termed the membrane potential. The membrane potential is utilized as the driving force of ATP synthase in the mitochondrial inner membrane. Although most of the electrons pass through the oxidative phosphorylation complexes, some occasionally leak out from it. The leaked electrons eventually produce reactive oxygen species (ROS). Therefore, mitochondria, especially oxidative phosphorylation complexes and mitochondrial DNA, are constitutively exposed to ROS and eventually disordered. It is believed that the damaged mitochondrial portion is selectively degraded by mitochondria autophagy (hereafter referred to as mitophagy), which serves to maintain mitochondrial function and eliminate the toxicity of ROS [1, 2].

In the last decade, several mitophagic pathways have been reported in mammalian cells, including PINK1-Parkin-dependent [3, 4], cellular differentiation-related [5–8], hypoxia-induced [9–11], and iron depletion-induced mitophagy pathways [12]. PINK1-Parkin-dependent mitophagy is the most commonly studied because this type of mitophagy can be easily induced through the loss of membrane potential in cultured mammalian cells. In this process, most of the mitochondria are degraded by autophagy up to 48 h; thus, mitophagy can be assessed by a decrease of mitochondrial proteins or DNA by immunoblotting, immunofluorescence microscopy, or quantitative polymerase chain reaction (PCR) [13–16]. Similarly, mitochondria elimination during the differentiation of erythroid cells can be detected by a decrease in mitochondrial proteins [5, 6]. In contrast to these relatively drastic mitochondria degradation pathways, the flux of hypoxia-induced or iron depletion-induced mitophagy is relatively low and, thus, difficult to detect by monitoring decreases in mitochondrial mass.

Recently, several groups, including ours, have reported that this low flux of mitophagy can be detected by mitochondria targeting fluorescent protein Keima (hereafter referred to as mito-Keima) [9, 17, 18]. Keima is an atypical fluorescent protein possessing two useful features for the detection of mitophagy [17, 19, 20]. One is that Keima is resistant to lysosomal proteases; thus, the delivery of mito-Keima into the lysosomal lumen can be monitored cumulatively upon mitophagy. The other is that Keima has a bimodal peak excitation wavelength of 440-nm and 586-nm light under neutral and acidic conditions, respectively, because of conformational changes of the chromophore responding to environmental pH [17]. Accordingly, using the mito-Keima system, we can detect mitochondria delivered into lysosomes as fluorescent signals selectively excited by 586-nm light [9]. These features enable us to detect low levels of mitophagy such as hypoxia-induced or iron depletion-induced mitophagy more sensitively (Fig. 1).

2 Materials

2.1 Mammalian Cells (See Note 1)

1. HeLa cells.
2. SH-SY5Y cells.
3. Mouse embryonic fibroblast (MEF) cells.
4. Platinum-E (Plat-E) Retroviral Packaging Cells, ecotropic (Cell Biolabs, Inc.).

2.2 Media for Cell Culture

1. DMEM high glucose (Wako).
2. Fetal bovine serum (FBS) (GIBCO).

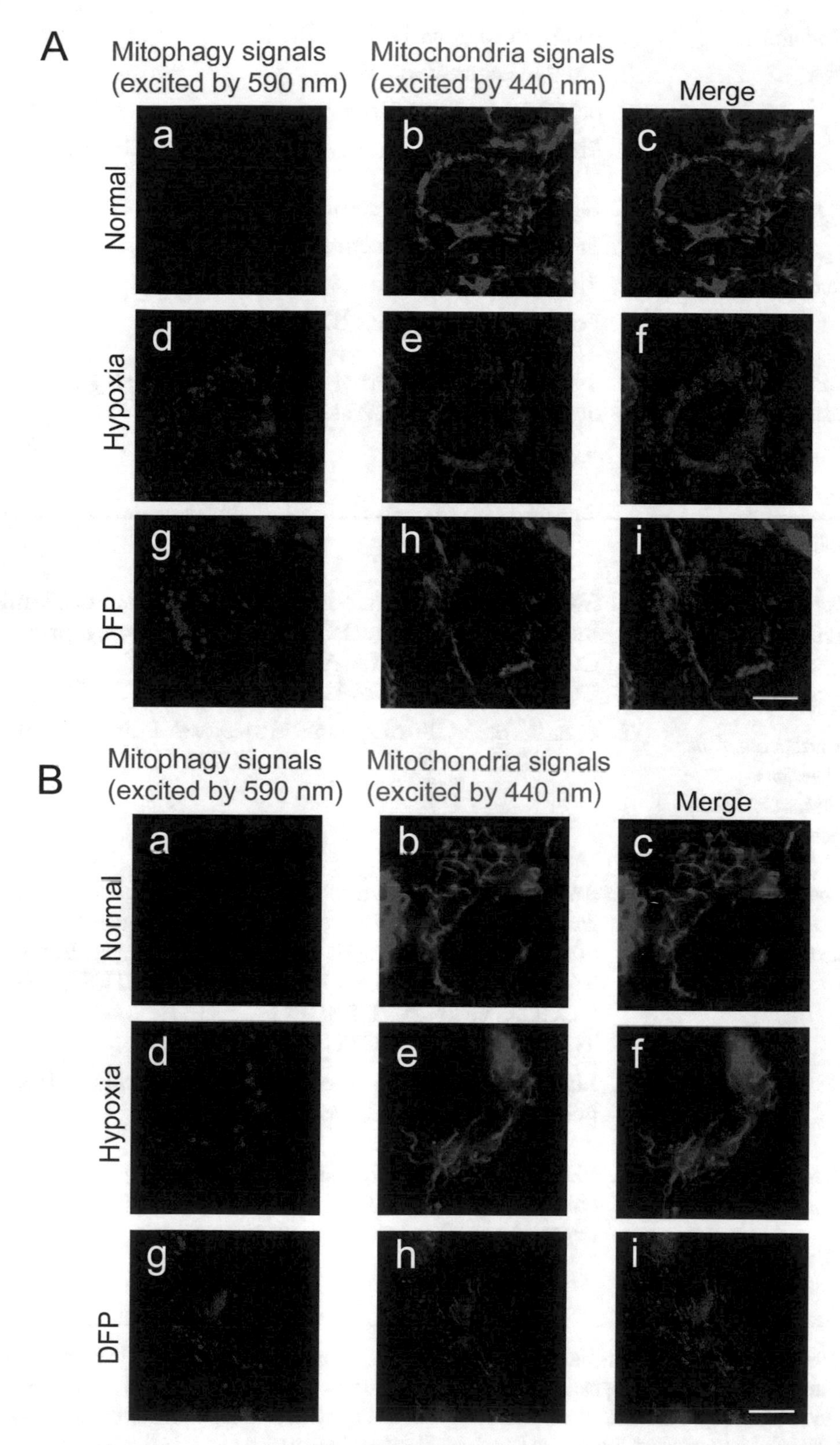

Fig. 1 Detecting hypoxia-induced and iron depletion-induced mitophagy. Stable HeLa (**A**) and SH-SY5Y (**B**) cell lines expressing mito-Keima were cultured under hypoxic conditions, DFP (an iron-chelating agent) treatment,

2.3 Plasmids and cDNA

1. pMXs-Puro (Cell Biolabs, Inc.).
2. pMT-mKeima-Red (MBL).
3. pcDNA3.1-Hygro (Invitrogen).
4. First-strand cDNA from mouse brain (BioChain).

2.4 Reagents

2.4.1 For Generation of Mito-Keima-Expressed Cells

1. Opti-MEM reduced serum medium (GIBCO).
2. FuGENE HD transfection reagent (Promega).
3. Puromycin (Wako).
4. Polybrene (Santa Cruz Biotechnology).

2.4.2 For Induction of Iron Depletion-Induced Mitophagy

1. 3-Hydroxy-1,2-dimethyl-4(1H)-pyridone [also referred to as deferiprone (DFP)] (Wako).

3 Methods

3.1 Generation of Stable Cell Lines Expressing Mito-Keima (See Note 2)

3.1.1 Construction of the Retroviral Vector for Mito-Keima Expression (pMXs-Puro-Mito-Keima)

1. Amplify the cDNA encoding mito-Keima from pMT-mKeima-Red as a template by PCR with the following primers: Fw: CGGGATCCGCCACCATGCTGAGCCTG, Rv: GGAATTCTTAACCGAGCAAAGAGTGGCGTG.
2. Ligate the PCR fragment into *Bam*HI-*Eco*RI sites of the pMXs-Puro vector.

3.1.2 Construction of mCAT1 Expression Vector (pcDNA3.1-Hygro-mCAT1)

1. Amplify the cDNA encoding the mouse cationic amino acid transporter (mCAT1) gene from the mouse first stranded cDNA pool by PCR with the following primers: Fw: CTAGCTAGCGCCACCATGGGCTGCAAAAACCTGCTCGG, Rv: CGGGATCCTCATTTGCAC TGGTCCAAGTTGCTGTCAGG.
2. Ligate the PCR fragment into *Nhe*I-*Bam*HI sites of the pcDNA3.1-Hygro vector.

3.1.3 Preparation of Retroviral Supernatant

1. Plate Plat-E cells at a density of 3×10^5 cells per well in a 6-well culture plate in 2 mL of DMEM with 10% FBS and culture overnight.

Fig. 1 (continued) or normal conditions. Mito-Keima signals excited by 590-nm light indicate mitochondria delivery into lysosomes by mitophagy (a, d, and g indicated as *red* in merged views). Mito-Keima signals excited by 440-nm light indicate mitochondria present in the cytoplasm (b, e, and h indicated as *green* in merged views). Merged views are shown in c, f, and i. Scale bar, 10 μm

2. Add 3 μg of pMXs-Puro-mito-Keima into a 1.5-mL microcentrifuge tube with 150 μL of Opti-MEM. Then, add 9 μL of FuGENE HD, mix well by pipetting gently, and incubate for 5 min at room temperature. The following process should be done under the biological safety hood level 2.
3. Add the plasmid-FuGENE HD mixture to Plat-E cells pre-cultured in 6-well plates and culture for 24 h.
4. Change the medium to 2 mL of DMEM with 10% FBS and incubate for another 24 h.
5. Collect the supernatant, which should contain retrovirus, and filtrate it with a 0.45 μm pore syringe filter to remove cell debris.
6. Add polybrene at a final concentration of 8 μg/mL, resulting in a retrovirus harboring the mito-Keima expression gene (mito-Keima retroviral solution). This retroviral solution can be stored at −80 °C for several months. However, we recommend using this solution immediately to prevent the reduction of titer.

3.1.4 Infection of Host Cells with the Mito-Keima Retroviral Solution

1. Here, we show the protocol using an ecotropic retroviral infection system to generate mito-Keima-expressed cell lines. Because HeLa and SH-SY5Y cells do not express the receptor for an ecotropic retrovirus, mCAT1 must be expressed in the cells prior to retroviral infection. Cells derived from mouse or rat, including MEF, intrinsically express mCAT1.
2. Plate HeLa or SH-SY5Y cells at a density of 5×10^4 cells per well in a 12-well plate in 1 mL of DMEM with 10% FBS and culture overnight. Prepare the pcDNA3.1-Hygro-mCAT1 vector-FuGENE HD mixture (1 μg of vector and 3 μL of FuGENE HD in 50 μL of Opti-MEM; *see* Section 3.1.3) and incubate for 5 min at room temperature. Add the vector-FuGENE HD mixture to the cells pre-cultured in 12-well plates and culture for 24 h.
3. Plate the host cells (such as MEF, or the pcDNA3.1-Hygro-mCAT1 vector-transfected HeLa, or SH-SY5Y cells) at a density of 1×10^5 cells per well in a 12-well culture plate in 1 mL of DMEM with 10% FBS and culture overnight.
4. Replace the medium to mito-Keima retroviral solution and culture for 24 h.
5. Replace the retroviral solution to 1 mL of DMEM with 1 μg/mL puromycin and culture for >3 days to exclude noninfected cells.
6. Maintain the mito-Keima-expressed cells with DMEM containing 1 μg/mL puromycin for continuous use in the following Section 3.1.5.

3.1.5 Cloning the Cells Expressing Mito-Keima with Moderate Expression Level

1. Plate the enriched cells expressing mito-Keima (Section 3.1.4) at a density of 0.4 cells per well in a 96-well culture plate with 0.2 mL of DMEM with 10% FBS and culture for 1–2 weeks.
2. Transfer the cells grown from a single colony at a density of 1×10^4 cells per well into a 96-well glass-bottom plate and culture overnight.
3. Observe the cells using a fluorescence microscope with the filter set for detecting Keima (*see* Section 3.2.1) and select the clone with moderate expression level. If the cells are cultured under normal conditions and not grown to high confluence, typical tubular forms of mitochondria should be observed when excited by 440 nm light; there should be no or only faint signals when excited by 590 nm light (*see* Fig. 1A and B, a and b).

3.2 Induction and Investigation of Mitophagy

*3.2.1 Hypoxia-Induced Mitophagy (See **Note 3**)*

1. Plate the stable cell line expressing mito-Keima (from Section 3.1.5) at a density of 1×10^4 cells per well into a 96-well glass-bottom plate in 100 μL of DMEM with 10% FBS and culture overnight (*see* **Note 4**).
2. Transfer the plate to a hypoxic chamber and culture for 24 h (*see* **Note 5**).
3. Observe the cells using a fluorescence microscope (Fig. 1). Mito-Keima is observed by fluorescence microscopy with the specific filter set for detecting the two excitation peaks of Keima (*see* **Note 6**). Mito-Keima signals excited by 590 or 440 nm light indicate mitochondria delivered into lysosomes by mitophagy (can be seen as punctate structures) or mitochondria present in the cytoplasm (can be seen as long or short tubular structure), respectively.

3.2.2 Iron Depletion-Induced Mitophagy

1. Plate the mito-Keima-expressed cells at a density of 1×10^4 cells per well into a 96-well glass-bottom plate in 100 μL of DMEM with 10% FBS and culture overnight (*see* **Note 4**).
2. Change the medium to 100 μL of DMEM with 10% FBS containing 1 mM DFP and culture the cells for 24 h.
3. Observe the cells directly with a fluorescence microscope (*see* Section 3.2.1, Fig. 1).

3.3 Estimate the Level of Mitophagy

1. The level of induced mitophagy can be estimated by counting the number of punctate structures observed when excited by 590-nm light.
2. Here, we show the method using Metamorph software to calculate mitophagy (*see* **Note 7**).

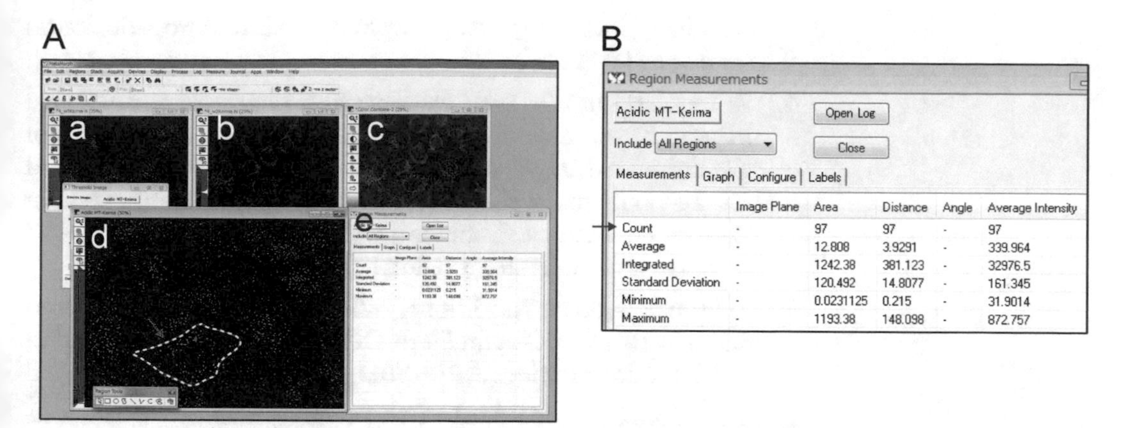

Fig. 2 Analysis of mitophagy signals with Metamorph. (**A**) Images, including mitophagy signals, excited by 590-nm light (a), mitochondria excited by 440-nm light (b), and merged view images (c) were processed by Metamorph software. The threshold image of mitophagy signals (d) was analyzed by the region measurement program. The summary window (e) showed several parameters of the cells indicated by *yellow arrows* in (d). (**B**) Enlarged image (e) showing parameters, including the number of mitophagy dots in the cells indicated by *red arrows*

3. Set the threshold and run the region measurements program to count mitophagy dots per cell (Fig. 2).
4. Calculate the percentage of cells with over 15 mitophagy dots as mitophagy positive from at least 50 cells.

4 Notes

1. We have tested HeLa, SH-SY5Y, and MEF cells using this method. In theory, most mammalian cultured cells should be able to be utilized. Recently, mito-Keima was used for detection of mitophagy in mice [21]. According to the analysis of mito-Keima transgenic mice, mitophagy can be detected in most type of cells using mito-Keima.
2. Cells transiently expressing mito-Keima can be used for examining mitophagy. However, when mito-Keima-expressed in the cells is either extremely strong or weak, mitophagy may not be detected with good signal-to-noise ratios, and, in mito-Keima strongly-expressed cells, occasional false-positive signals may be detected. Hence, we strongly recommend generating stable cell lines expressing mito-Keima and to clone only the cells with moderate expression levels.
3. This method, using mito-Keima, is very sensitive and, thus, useful for both Parkin-dependent and -independent

mitophagy, including mitophagy in hypoxia and iron-depletion conditions [9, 12].

4. We regularly use 96-well glass-bottom plates for mito-Keima analysis. In the case of mito-Keima-expressed HeLa cells, the cells are plated to 96-well glass-bottom plates at a density of 1×10^4 cells per well and cultured overnight until reaching 70–80% confluency. Differently sized glass-bottom dishes/plates are also available, if cellular confluency is adjusted.
5. To induce hypoxic conditions, cells were cultured under conditions with 1% of O_2 and 5% of CO_2 using a multi-gas incubator (Astec corporation: APM-30D).
6. Mito-Keima is observed with a fluorescence microscope equipped with the specific filter set for detecting the two excitation peaks of Keima. Mito-Keima localizes to the mitochondrial matrix where the neutral environment is excited by the 430 ± 24-nm excitation filter (Chroma), whereas mito-Keima delivered into the lysosomal lumen is excited by the 560 ± 40-nm excitation filter (Chroma). Both excitations are detected with the 624 ± 20-nm emission filter (Chroma) and a dichroic mirror for Texas Red (Semrock).
7. The estimation of mitophagy level can be performed using Metamorph or other imaging software, including ImageJ (Dr. Wayne Rasband).

Acknowledgment

This work was supported in part by the Japan Society for the Promotion of Science KAKENHI Grant numbers 26291039 (T.K.), 16H01198 (T.K.), 16H01384 (T.K.), 15 K18501 (S.Y.), Yujin Memorial Grant (Niigata University School of Medicine) (T.K.), The Sumitomo Foundation (T.K.), Astellas Foundation for Research on Metabolic Disorders (T.K.), and Takeda Science Foundation (S.Y. and T.K.).

References

1. Twig G, Elorza A, Molina AJ, Mohamed H, Wikstrom JD, Walzer G, Stiles L, Haigh SE, Katz S, Las G, Alroy J, Wu M, Py BF, Yuan J, Deeney JT, Corkey BE, Shirihai OS (2008) Fission and selective fusion govern mitochondrial segregation and elimination by autophagy. EMBO J 27(2):433–446. doi:10.1038/sj.emboj.7601963
2. Wallace DC (2005) A mitochondrial paradigm of metabolic and degenerative diseases, aging, and cancer: a dawn for evolutionary medicine. Annu Rev Genet 39:359–407. doi:10.1146/annurev.genet.39.110304.095751
3. Geisler S, Holmstrom KM, Skujat D, Fiesel FC, Rothfuss OC, Kahle PJ, Springer W (2010) PINK1/Parkin-mediated mitophagy is dependent on VDAC1 and p62/SQSTM1. Nat Cell Biol 12(2):119–131. doi:10.1038/ncb2012
4. Matsuda N, Sato S, Shiba K, Okatsu K, Saisho K, Gautier CA, Sou YS, Saiki S, Kawajiri S, Sato F, Kimura M, Komatsu M, Hattori N, Tanaka

K (2010) PINK1 stabilized by mitochondrial depolarization recruits Parkin to damaged mitochondria and activates latent Parkin for mitophagy. J Cell Biol 189(2):211–221. doi:10.1083/jcb.200910140

5. Novak I, Kirkin V, McEwan DG, Zhang J, Wild P, Rozenknop A, Rogov V, Lohr F, Popovic D, Occhipinti A, Reichert AS, Terzic J, Dotsch V, Ney PA, Dikic I (2010) Nix is a selective autophagy receptor for mitochondrial clearance. EMBO Rep 11(1):45–51. doi:10.1038/embor.2009.256
6. Sandoval H, Thiagarajan P, Dasgupta SK, Schumacher A, Prchal JT, Chen M, Wang J (2008) Essential role for Nix in autophagic maturation of erythroid cells. Nature 454 (7201):232–235. doi:10.1038/nature07006
7. Schweers RL, Zhang J, Randall MS, Loyd MR, Li W, Dorsey FC, Kundu M, Opferman JT, Cleveland JL, Miller JL, Ney PA (2007) NIX is required for programmed mitochondrial clearance during reticulocyte maturation. Proc Natl Acad Sci U S A 104(49):19500–19505. doi:10.1073/pnas.0708818104
8. Zhang Y, Goldman S, Baerga R, Zhao Y, Komatsu M, Jin S (2009) Adipose-specific deletion of autophagy-related gene 7 (atg7) in mice reveals a role in adipogenesis. Proc Natl Acad Sci U S A 106(47):19860–19865. doi:10.1073/pnas.0906048106
9. Hirota Y, S-i Y, Kurihara Y, Jin X, Aihara M, Saigusa T, Kang D, Kanki T (2015) Mitophagy is primarily due to alternative autophagy and requires the MAPK1 and MAPK14 signaling pathways. Autophagy 11(2):332–343. doi:10.1080/15548627.2015.1023047
10. Liu L, Feng D, Chen G, Chen M, Zheng Q, Song P, Ma Q, Zhu C, Wang R, Qi W, Huang L, Xue P, Li B, Wang X, Jin H, Wang J, Yang F, Liu P, Zhu Y, Sui S, Chen Q (2012) Mitochondrial outer-membrane protein FUNDC1 mediates hypoxia-induced mitophagy in mammalian cells. Nat Cell Biol 14 (2):177–185. doi:10.1038/ncb2422
11. Zhang H, Bosch-Marce M, Shimoda LA, Tan YS, Baek JH, Wesley JB, Gonzalez FJ, Semenza GL (2008) Mitochondrial autophagy is an HIF-1-dependent adaptive metabolic response to hypoxia. J Biol Chem 283 (16):10892–10903. doi:10.1074/jbc.M800102200
12. Allen GF, Toth R, James J, Ganley IG (2013) Loss of iron triggers PINK1/Parkin-independent mitophagy. EMBO Rep 14 (12):1127–1135. doi:10.1038/embor.2013.168
13. Narendra D, Kane LA, Hauser DN, Fearnley IM, Youle RJ (2010) p62/SQSTM1 is required for Parkin-induced mitochondrial clustering but not mitophagy; VDAC1 is dispensable for both. Autophagy 6 (8):1090–1106
14. Narendra D, Tanaka A, Suen DF, Youle RJ (2008) Parkin is recruited selectively to impaired mitochondria and promotes their autophagy. J Cell Biol 183(5):795–803. doi:10.1083/jcb.200809125
15. Vives-Bauza C, Zhou C, Huang Y, Cui M, de Vries RL, Kim J, May J, Tocilescu MA, Liu W, Ko HS, Magrane J, Moore DJ, Dawson VL, Grailhe R, Dawson TM, Li C, Tieu K, Przedborski S (2010) PINK1-dependent recruitment of Parkin to mitochondria in mitophagy. Proc Natl Acad Sci U S A 107(1):378–383. doi:10.1073/pnas.0911187107
16. Lazarou M, Sliter DA, Kane LA, Sarraf SA, Wang C, Burman JL, Sideris DP, Fogel AI, Youle RJ (2015) The ubiquitin kinase PINK1 recruits autophagy receptors to induce mitophagy. Nature 524(7565):309–314. doi:10.1038/nature14893
17. Katayama H, Kogure T, Mizushima N, Yoshimori T, Miyawaki A (2011) A sensitive and quantitative technique for detecting autophagic events based on lysosomal delivery. Chem Biol 18(8):1042–1052. doi:10.1016/j.chembiol.2011.05.013
18. Kageyama Y, Hoshijima M, Seo K, Bedja D, Sysa-Shah P, Andrabi SA, Chen W, Hoke A, Dawson VL, Dawson TM, Gabrielson K, Kass DA, Iijima M, Sesaki H (2014) Parkin-independent mitophagy requires Drp1 and maintains the integrity of mammalian heart and brain. EMBO J 33(23):2798–2813. doi:10.15252/embj.201488658
19. Kogure T, Karasawa S, Araki T, Saito K, Kinjo M, Miyawaki A (2006) A fluorescent variant of a protein from the stony coral Montipora facilitates dual-color single-laser fluorescence cross-correlation spectroscopy. Nat Biotechnol 24 (5):577–581. doi:10.1038/nbt1207
20. Violot S, Carpentier P, Blanchoin L, Bourgeois D (2009) Reverse pH-dependence of chromophore protonation explains the large stokes shift of the red fluorescent protein mKeima. J Am Chem Soc 131(30):10356–10357. doi:10.1021/ja903695n
21. Sun N, Yun J, Liu J, Malide D, Liu C, Rovira II, Holmstrom KM, Fergusson MM, Yoo YH, Combs CA, Finkel T (2015) Measuring in vivo mitophagy. Mol Cell 60(4):685–696. doi:10.1016/j.molcel.2015.10.009

Methods in Molecular Biology (2018) 1759: 151–160
DOI 10.1007/7651_2017_18

Published online: 22 March 2017

Monitoring Mitophagy During Aging in *Caenorhabditis elegans*

Nikolaos Charmpilas*, Konstantinos Kounakis*, and Nektarios Tavernarakis

Abstract

Mitochondria constitute the main energy-producing centers of eukaryotic cells. In addition, they are involved in several crucial cellular processes, such as lipid metabolism, calcium buffering, and apoptosis. As such, their malfunction can be detrimental for proper cellular physiology and homeostasis. Mitophagy is a mechanism that protects and maintains cellular function by sequestering harmful or dysfunctional mitochondria to lysosomes for degradation. In this report, we present experimental procedures for quantitative, in vivo monitoring of mitophagy events in the nematode *Caenorhabditis elegans*.

Keywords: Aging, Autophagy, *Caenorhabditis elegans*, Mitochondria, Mitophagy, Ratiometric imaging, Rosella biosensor

1 Introduction

The nematode *C. elegans* is widely appreciated as an invaluable model for the study of fundamental cellular processes. Its versatility can be attributed to a unique combination of beneficial properties the organism possesses. First and foremost, its small and transparent body renders it ideal for in vivo, noninvasive, and real-time monitoring through simple optical and fluorescent microscopy. Furthermore, the organism can be manipulated genetically with relative ease, thus facilitating the dissection of molecular pathways through forward and reverse genetics. This endeavor is further supported by the invariant number and lineage of the organism's cells and the complete knowledge of its neuronal connectome. Finally, its short lifespan and its abundant reproduction make it ideal for the study of processes related to aging [1].

*The authors Nikolaos Charmpilas and Konstantinos Kounakis contributed equally to this chapter.

Organelle quality and quantity can be crucial for proper cellular function. Cells need to tightly regulate the amount of each organelle so that they can meet their current needs without excess, while damaged organelles need to be promptly removed and replaced. Autophagy is a cellular mechanism which contributes greatly to both these goals by deconstructing unwanted organelles. There is evidence for the existence of specific autophagic pathways for most cellular components, including peroxisomes [2, 3], lipid droplets [4], lysosomes [5], the nucleus [6], ER [7], ribosomes [8], and mitochondria.

Mitophagy, the specific macroautophagic mechanism targeting mitochondria, plays a critical role in cellular survival. Respiration causes the formation of reactive oxygen species (ROS) that can damage vital macromolecules and especially the vulnerable mitochondrial DNA. This damage can cause anomalies in mitochondrial function and render the malfunctioning organelles dangerous for the whole cell. Mitophagy removes such aberrant organelles, protecting cells and maintaining normal organismal function and lifespan [9–11]. Impaired mitochondrial function has been implicated in the onset of aging [12] and severe disorders, such as neurodegeneration [13] and cardiomyopathies [14]. Dysfunctions in mitophagy have been also associated with Parkinson's disease [15, 16].

The mechanisms which initiate mitophagy exhibit some variation among organisms, since not all species use the same signaling to recruit mitochondria into autophagosomes [17]. Yeast possesses a unique pathway involving Atg-32 and Atg-11, while other eukaryotes generally use the NIX (NIP3-like protein X), FUNDC1, and PINK1/PARKIN pathways [17, 18]. The Bcl-2 family protein NIX is a receptor that has been associated with mitophagy during cell maturation events [19, 20]. It has also been shown to facilitate regular mitochondrial turnover based on organelle energetic status in collaboration with the small GTPase Rheb [21]. FUNDC1 is an outer mitochondrial membrane protein that is necessary for hypoxia-induced mitophagy in mammalian systems [22]. PINK1 is a kinase that normally enters the mitochondria and is deactivated by PARL (presenilin-associated rhomboid-like protease) but becomes stabilized in the outer mitochondrial membrane of dysfunctional organelles, thus recruiting the ubiquitin ligase Parkin. The combination of ubiquitination and phosphorylation by this newly formed complex leads to the labeling of proteins on the outer mitochondrial membrane and the recruitment of the defunct mitochondria into autophagosomes (Fig. 1a). Regardless of how mitophagy is initiated, all pathways converge into a similar sequence of events. Mitochondria are enclosed by the isolation membrane of the autophagosomes, which binds Atg-8/LC3 (LGG-1 in *C. elegans*). The autophagosomes are eventually sequestered to lysosomes, where mitochondria are degraded [17, 23].

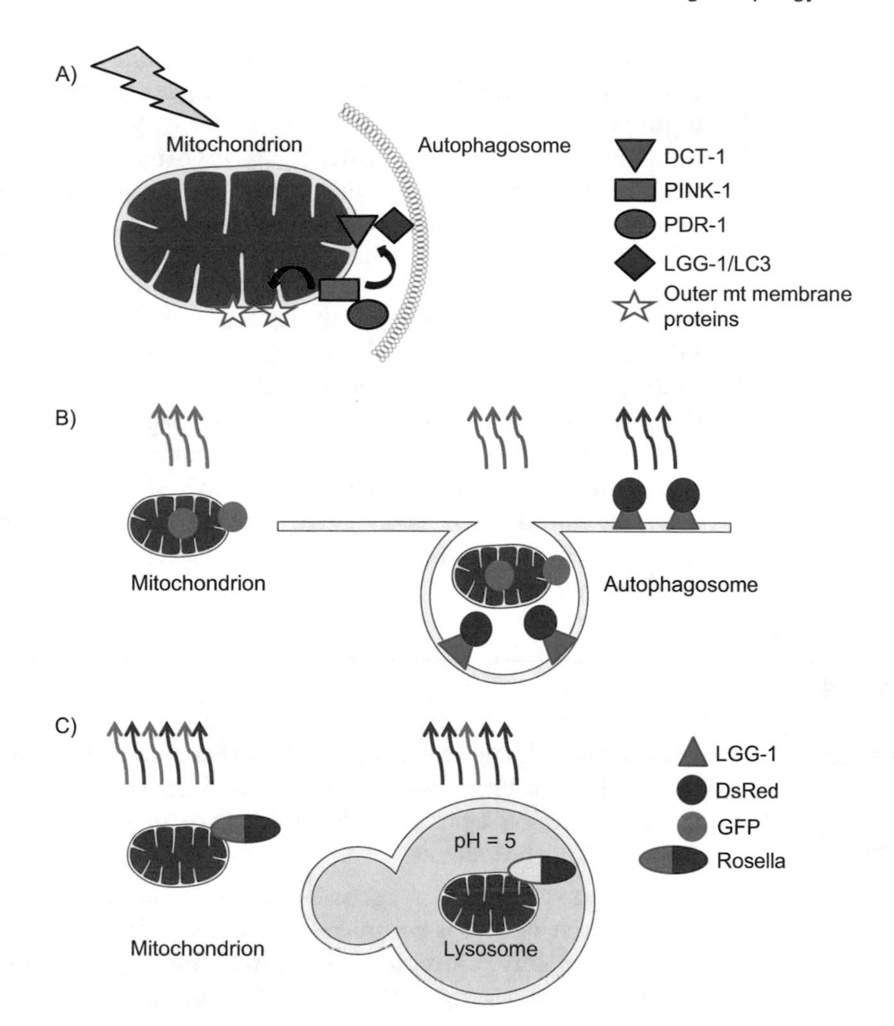

Fig. 1 A *C. elegans* platform for monitoring mitophagy in vivo. (**a**) Mitophagy components in *C. elegans*. When the mitochondrial environment is challenged by exogenous or endogenous insults, mitochondria-specific macroautophagy (mitophagy) is activated to prevent accumulation of dysfunctional organelles. In *C. elegans*, mitophagy is mediated by DCT-1, a receptor which localizes on the outer mitochondrial membrane, as well as the PINK-1 kinase and PDR-1 E3 ubiquitin ligase. Combined ubiquitination and phosphorylation of several outer mitochondrial membrane proteins, among them DCT-1 itself, by the PINK-1/PDR-1 complex signals for the sequestration of mitochondria to autophagosomes and finally to lysosomes for degradation. (**b**) Mitophagy can be monitored by observing the co-localization of GFP-tagged mitochondria with DsRed-tagged autophagosomes. (**c**) Alternatively, mitophagy events can be quantified by observing the reduction of green channel emission of the hybrid fluorescent reporter Rosella upon entry to the acidic lysosomes

Initial evidence for the existence of active mitophagy in nematodes arose from reports supporting that paternal sperm mitochondria are eliminated upon fertilization through macroautophagy, thus preventing aberrant heteroplasmy of the mitochondrial genome [24, 25]. Recently, our lab identified DCT-1 as the putative worm homologue of the mammalian mitophagy receptor NIX/BNIP-3 [26]. DCT-1 acts cooperatively with PINK-1 kinase and Parkin E3 ubiquitin ligase to orchestrate mitophagy induction under stressful conditions (Fig. 1a). Intriguingly, our work also highlighted that mitophagy accounts for the majority of autophagic influx in response to stimuli which are reported to trigger autophagy. Furthermore, intact mitophagy is indispensable for the extended lifespan of various long-lived mutants, such as animals with reduced insulin/IGFR signaling. Mitophagy is also essential for the extended longevity of worms which exhibit perturbed electron transport chain activity or bear a depletion of frataxin, a mitochondrial protein involved in iron-sulfur cluster biogenesis. Frataxin inhibition causes mitochondrial stress due to iron starvation, an insult which strongly induces mitophagy [27]. Collectively, mitophagy serves as a core cellular mechanism, which preserves homeostasis and determines lifespan in cases where the mitochondrial environment is challenged. In this article, we describe the experimental methodology primarily used in our lab to track mitophagy events in *C. elegans* in vivo.

2 Materials

2.1 C. elegans Manipulation

2.1.1 Equipment

1. Worm pick and eyelash hair: We use a platinum pick for routinely transferring worms in plates and an eyelash hair to transfer them on agarose pads just before observation to avoid stressing the animals.
2. 2% agarose pads: 1 g agarose powder in 50 mL distilled water and heat until it is completely dissolved. While the solution is still hot, leave a droplet (~1 cm in diameter) on top of a glass slide. Put a second slide on top and press until a large gel pad is formed.
3. Dissecting microscope.

2.1.2 Solutions and Chemicals

1. M9 buffer: Dissolve 3 g KH_2PO_4, 6 g Na_2HPO_4, and 5 g NaCl in 1 L distilled water. Autoclave and add 1 mL 1 M $MgSO_4$. M9 buffer can be stored at room temperature or 4 °C.
2. Tetramisole: Tetramisole is used for the anesthetization of the animals. Prepare a 100 mM tetramisole solution by dissolving 1.2 g tetramisole powder in 50 mL M9 solution, and store at 4 °C. We normally dilute this further and prepare a 20 mM working solution for everyday use.

2.2 Nematode Food

1. OP50 bacterial food: Inoculate 50 mL of liquid, sterile LB broth with a single OP50 colony, and let it grow for approximately 8 h at 37 °C with shaking. Plate 180–200 μL of the OP50 culture in the center of freshly prepared NGM plates, and let the bacterial lawn grow overnight (for approximately

16 h). In the next morning, the plates are ready to be used for animal transferring. This procedure requires sterile conditions.

2. Petri dishes containing nematode growth medium (NGM): For preparing 1 L of NGM, dissolve 3 g NaCl, 2.5 g bacto-peptone, 17 g agar, and 0.2 g streptomycin sulfate powders in a flask containing 900 mL distilled H_2O. Stir the mix for 5–10 min and then autoclave. Let the flask cool down to 55 °C, and add 1 mL $CaCl_2$ (stock solution 1 M), 1 mL $MgSO_4$ (stock solution 1 M), 1 mL cholesterol (stock solution 5 mg/mL), and 1 mL nystatin (stock solution 10 mg/mL). Finally, add 25 mL 1 M phosphate KPO_4 buffer (pH = 6.0) and distilled water to the final volume of 1 L. Stir the mix for 5–10 min and dispense the medium into the plates in the desired volume.
3. RNAi Petri dishes: The preparation of those plates is similar to the one described above for NGM plates. However, instead of adding streptomycin sulfate powder before autoclaving, in this case add ampicillin after autoclaving, at a final concentration of 100 μg/mL.
4. Petri dishes containing RNA inhibition medium: To perform gene silencing experiments, streak the desired HT115 bacterial strain containing the RNAi expressing vector of choice (a pL4440 vector backbone containing a fragment of your gene of interest) on LB agar plates with ampicillin at a final concentration of 100 μg/mL and tetracycline at a final concentration of 10 μg/mL. Incubate the plates at 37° for 16–18 h. Inoculate 5 mL of liquid LB with ampicillin and tetracycline (at similar concentrations with those of solid plates) with a single colony from the plate. Incubate at 37° for 16–18 h with shaking. Inoculate the desired amount of LB containing ampicillin with the bacterial culture, at a ratio of 50 μL overnight culture per 1 mL of LB with ampicillin. Incubate for 2–4 h at 37° with shaking. Disperse 200 μL of a mix of the over-day culture plus IPTG at a final concentration of 2 mM at the center of freshly prepared RNAi Petri dishes. Incubate overnight at room temperature to allow bacterial lawn grow. This procedure requires sterile conditions. In the next morning, animals can be transferred on these plates.

2.3 Chemical Compounds

1. Paraquat (N,N′-dimethyl-4,4′-bipyridinium dichloride): Prepare 0.5 M stock solution, and dilute it in distilled water, to a final concentration of 8 mM per plate. Paraquat stock solution can be stored at 4 °C. *Note*: Paraquat is extremely toxic for human beings; thus, any contact should be avoided!
2. Carbonyl cyanide m-chlorophenyl hydrazone (CCCP): Dissolve 100 mg CCCP in 10 mL dimethyl sulfoxide (DMSO), aliquot, and store at −20 °C. Add to the plates at a final concentration of 10 mM.

2.4 Nematode Strains

1. IR1284: WT;*Is*[$p_{myo\text{-}3}$mtGFP];*Ex*[$p_{lgg\text{-}1}$DsRed::LGG-1].
2. IR1511: WT;*Ex*[$p_{myo\text{-}3}$DsRed::LGG-1;$p_{dct\text{-}1}$DCT-1::GFP].
3. IR1631: WT;*Ex*[$p_{myo\text{-}3}$TOMM-20::Rosella].

3 Methods

3.1 CCCP and Paraquat Treatment

1. Add both paraquat and CCCP on NGM plates, at a final concentration of 8 and 10 mM per plate, respectively, on the top of UV-treated OP50 bacteria (*see* **Notes 1** and **2**). We normally kill bacteria through exposure to UV light for 15 min in a UV cross-linker.
2. Spread the drug-containing solution throughout the entire surface of the plate, and let it dry at room temperature.
3. Place at least 20 nematodes on each of the drug-containing plates as well as on plates with pure solvent (distilled water or DMSO) for negative control (*see* **Note 3**).
4. Monitor induction of mitophagy 2 days later using the methods described below.

3.2 Sample Preparation for Imaging (All Strains)

1. Place a 10–20 μL droplet of 20 mM tetramisole on a 2% agarose pad.
2. Use an eyelash hair to transfer the worms onto the droplet. Exposure of the worms to tetramisole will render them immobile within a few seconds and allow efficient imaging (*see* **Note 4**).
3. Cover gently with a coverslip.

Worms should not dry out during the imaging process. Ignore any worms that appear to have been damaged due to mishandling (worms with vulva rupture).

3.3 Monitoring Mitophagy by Simultaneous Labeling of Mitochondria and Autophagosomes

The first method for quantifying mitophagy relies on observing co-localization between the mitochondrial outer membrane, identified by GFP either fused to the DCT-1 mitophagy receptor or marking the mitochondrial matrix and the autophagosomal isolation membrane, identified by DsRed fused with LGG-1 (Fig. 1b) [26]. This method follows a similar principle to older techniques that utilized fluorescent antibodies and MitoTracker staining (*see* **Note 5**) [24]:

1. Photograph body wall muscle cells using a confocal fluorescent microscope (equivalent to a Zeiss AxioObserver Z1 confocal microscope) and performing *z*-stack imaging.
2. Use image analysis software of choice to count mitochondria engulfed by autophagosomes per body wall muscle cell. These can be observed as puncta where mitochondrially localized GFP (mtGFP) and the autophagosomal marker (LGG-1::DsRed) co-localize (*see* **Notes 6** and **7**).

3. Perform statistical analysis with appropriate software (Microsoft Excel/Graphpad Prism) to examine possible changes in mitophagy between samples. Use the student's t-test for comparison between samples with a significance cutoff level of $p < 0.05$. Use ANOVA corrected by the post hoc Bonferroni test for multiple comparisons. We recommend examining at least 50 individual cells from different animals per experiment, and repeating each experiment thrice.

3.4 Monitoring Mitophagy Using the Rosella Biosensor

As an alternative method to monitor mitophagy events, we have utilized the Rosella biosensor (Fig. 1c). Rosella is a dual fluorescent reporter which originated from the combination of a pH-sensitive GFP variant and a pH-insensitive DsRed variant [28]. We have fused Rosella with an amino terminal fragment of TOMM-20, which localizes in the outer mitochondrial membrane and mediates import of nucleus-encoded proteins into mitochondria [29]. Via this approach, we can quantify mitophagy events by merely assessing the fraction of green fluorescence which has quenched as a consequence of fusion of mitochondria-tagged Rosella with lysosomes (*see* **Note 8**). Mitochondria-tagged Rosella was specifically overexpressed in *C. elegans* body wall muscle cells, using the minor isoform of myosin heavy chain promoter (*myo-3* gene):

1. Observe the animals under a fluorescent microscope (equivalent to Zeiss Axio Imager Z2 Epifluorescence/DIC Micro scope), and acquire images of the head region of 20 or more animals using a 10X lens and a digital camera.
2. Using the ImageJ/FIJI image processing package, split the channels to separately quantify the green and red fluorescence intensity in a representative head region, just anterior to the animal's intestine to avoid gut autofluorescence (*see* **Notes 6**, **7**, and **9**).
3. Divide the individual values of green versus red fluorescence from single animals using Microsoft Excel or GraphPad Prism. Comparisons between samples can be performed using the student's t-test with a significance cutoff level of $p < 0.05$. Use ANOVA corrected by the post hoc Bonferroni test for multiple comparisons between different experimental conditions.

Under mitophagy-inducing conditions, the ratio of green versus red fluorescence is expected to drop significantly, as more mitochondria will be sequestered to acidic lysosomes for degradation. We recommend examining at least 50 individual animals per experiment and repeating each experiment thrice.

4 Notes

1. An important parameter which should be taken under consideration is that the measurements should be conducted in animals of the same age under the desired experimental conditions. Macroautophagy is inseparably linked with aging, and it can be influenced by interventions which accelerate or delay its onset [30]. For mitophagy measurements described here, animal synchronization based on the crescent-like vulval morphology of L4-staged worms is preferable than synchronization in the egg stage via hypochlorite treatment or egg laying. This is because treatment with CCCP or paraquat is restrictive for normal development and only few of the treated animals can reach adulthood.
2. Unless your study is explicitly related to starvation/caloric restriction, you should always only use non-starved animals in your experiments. We recommend ensuring your strains have been well fed for at least three generations to avoid any transgenerational effects that might be inherited from starved parents [31, 32].
3. We recommend preparing fresh solutions for both the mitophagy-inducing drugs (paraquat and CCCP) prior to each experiment. Freeze-thaw cycles may reduce drug efficiency and diminish the robust induction of mitophagy that makes them ideal positive controls.
4. Sodium azide, widely used for worm anesthesia before imaging, is a strong inhibitor of oxidative phosphorylation [33]. Since it is detrimental for mitochondrial function, it is also a potential activator of mitophagy. Hence, we suggest the exclusive use of tetramisole to anesthetize the animals prior to observation.
5. In the case of the autophagosome co-localization method, an increase in observed LGG-1::GFP puncta might not necessarily result from increased autophagy, but instead represent a side effect of a dysfunctional pathway. For instance, defunct autolysosomes can lead to autophagosome accumulation and increased LGG-1::GFP puncta [34]. Hence, it is advisable to also rely on an alternative approach for drawing safe conclusions regarding mitophagy induction.
6. Thoroughly document the method and specific parameters of the analysis process, and ensure consistency across experiments. This is particularly important if you use an automated method to count the amount of LGG-1::DsRed and mitochondrial GFP co-localization.

7. It is important to distinguish actual DsRed fluorescence from bleed-through emission of GFP into "red wavelengths." This can happen if you overexpose your samples during imaging. In the case of the LGG-1::DsRed and mitochondrial GFP co-localization method, observing perfect alignment between all green and red spots may be an indication of bleed-through.
8. As an alternative to the Rosella biosensor, we suggest the use of pHluorin [35]. This is a pH-sensitive version of GFP whose fluorescence is quenched in acidic compartments, such as the lysosomes. Tagging of an outer membrane mitochondrial protein with pHluorin would allow monitoring of mitophagy events based on a single-color method.
9. Always document the parameters of your imaging as thoroughly as possible, and ensure they remain consistent between experiments. This is particularly important for the Rosella method, as changes in the imaging process might not affect both color channels equally, thus leading to skewed fluorescence ratios.

Acknowledgments

Work in the authors' laboratory is funded by grants from the European Research Council (ERC), the European Commission Framework Programmes, and the Greek Ministry of Education. Konstantinos Kounakis is a recipient of an Onassis Foundation postgraduate scholarship.

References

1. Corsi AK, Wightman B, Chalfie M (2015) A transparent window into biology: a primer on *Caenorhabditis elegans*. Genetics 200 (2):387–407
2. Katarzyna ZR, Suresh S (2016) Autophagic degradation of peroxisomes in mammals. Biochem Soc Trans 44(2):431–440
3. Oku M, Sakai Y (2016) Pexophagy in yeasts. Biochim Biophys Acta 1863(5):992–998
4. Cingolani F, Czaja MJ (2016) Regulation and functions of autophagic lipolysis. Trends Endocrinol Metab 27(10):696–705
5. Kawabata T, Yoshimori T (2016) Beyond starvation: an update on the autophagic machinery and its functions. J Mol Cell Cardiol 95:2–10
6. Mijaljica D, Devenish RJ (2013) Nucleophagy at a glance. J Cell Sci 126(Pt 19):4325–4330
7. Khaminets A et al (2015) Regulation of endoplasmic reticulum turnover by selective autophagy. Nature 522(7556):354–358
8. Suzuki K (2013) Selective autophagy in budding yeast. Cell Death Differ 20(1):43–48
9. Lemasters JJ (2005) Selective mitochondrial autophagy, or mitophagy, as a targeted defense against oxidative stress, mitochondrial dysfunction, and aging. Rejuvenation Res 8:3–5
10. Bergamini E (2006) Autophagy: a cell repair mechanism that retards ageing and age-associated diseases and can be intensified pharmacologically. Mol Asp Med 27:403–410
11. Kim I, Rodriguez-Enriquez S, Lemasters JJ (2007) Selective degradation of mitochondria by mitophagy. Arch Biochem Biophys 462:245–253
12. Sun N, Youle RJ, Finkel T (2016) The mitochondrial basis of aging. Mol Cell 61:654–666
13. Redmann M, Darley-Usmar V, Zhang J (2016) The role of autophagy, mitophagy and lysosomal functions in modulating bioenergetics and survival in the context of redox and

proteotoxic damage: implications for neurodegenerative diseases. Aging Dis 7:150–162
14. Tong M, Sadoshima J (2016) Mitochondrial autophagy in cardiomyopathy. Curr Opin Genet Dev 38:8–15
15. Zhang J (2013) Autophagy and mitophagy in cellular damage control. Redox Biol 1:19–23
16. Voigt A, Berlemann LA, Winklhofer KF (2016) The mitochondrial kinase PINK1: functions beyond mitophagy. J Neurochem 139: 232–239. doi: 10.1111/jnc.13655
17. Youle RJ, Narendra DP (2011) Mechanisms of mitophagy. Nat Rev Mol Cell Biol 12:9–14
18. Yamaguchi O et al (2016) Receptor-mediated mitophagy. J Mol Cell Cardiol 95:50–56
19. Schweers RL et al (2007) NIX is required for programmed mitochondrial clearance during reticulocyte maturation. Proc Natl Acad Sci U S A 104(p):19500–19505
20. Sandoval H et al (2008) Essential role for Nix in autophagic maturation of erythroid cells. Nature 454:232–235
21. Melser S et al (2013) Rheb regulates mitophagy induced by mitochondrial energetic status. Cell Metab 17:719–730
22. Liu L et al (2012) Mitochondrial outer-membrane protein FUNDC1 mediates hypoxia-induced mitophagy in mammalian cells. Nat Cell Biol 14:177–185
23. Palmisano NJ and Meléndez A. (2016) Detection of autophagy in *Caenorhabditis elegans*. Cold Spring Harbor Protoc 2016(2):pdb. top070466
24. Al Rawi S et al (2011) Postfertilization autophagy of sperm organelles prevents paternal mitochondrial DNA transmission. Science 334(6059):1144–1147
25. Sato M, Sato K (2011) Degradation of paternal mitochondria by fertilization-triggered autophagy in *C. elegans* embryos. Science 334 (6059):1141–1144
26. Palikaras K, Lionaki E, Tavernarakis N (2015) Coordination of mitophagy and mitochondrial biogenesis during ageing in *C. elegans*. Nature 521(7553):525–528
27. Schiavi A et al (2015) Iron-starvation-induced mitophagy mediates lifespan extension upon mitochondrial stress in *C. elegans*. Curr Biol 25(14):1810–1822
28. Rosado CJ et al (2008) Rosella: a fluorescent pH-biosensor for reporting vacuolar turnover of cytosol and organelles in yeast. Autophagy 4:205–213
29. Dudek J, Rehling P, van der Laan M (2013) Mitochondrial protein import: common principles and physiological networks. Biochim Biophys Acta 1833(2):274–285
30. Rubinsztein DC, Mariño G, Kroemer G (2011) Autophagy and aging. Cell 146 (5):682–695
31. Rechavi O et al (2014) Starvation-induced transgenerational inheritance of small RNAs in *C. elegans*. Cell 158:277–287
32. Jobson MA et al (2015) Transgenerational effects of early life starvation on growth, reproduction and stress resistance in *C. elegans*. Genetics 201:1–37
33. Bogucka K, Wojtczak L (1966) Effect of sodium azide on oxidation and phosphorylation processes in rat-liver mitochondria. Biochim Biophys Acta 122(3):381–392
34. Zhang H et al (2015) Guidelines for monitoring autophagy in *Caenorhabditis elegans*. Autophagy 11:9–27
35. Kavalali ET, Jorgensen EM (2014) Visualizing presynaptic function. Nat Neurosci 17 (1):10–16

Methods in Molecular Biology (2018) 1759: 161–172
DOI 10.1007/7651_2017_40

Published online: 10 May 2017

Monitoring of Iron Depletion-Induced Mitophagy in Pathogenic Yeast

Koichi Tanabe and Minoru Nagi

Abstract

Mitophagy, which is the degradation of mitochondria via selective autophagic machinery, is thought to be involved in regulating the mass and function of mitochondria. Methods for detection of mitophagy have been reported for several fungal cells including some budding yeast, methylotrophic yeast, and filamentous fungi. Mitophagy in *Saccharomyces cerevisiae* is activated under nitrogen-poor conditions; however, the regulatory mechanism of mitophagy in most fungi has not been elucidated. Here we describe methods to monitor mitophagy in the pathogenic yeast *Candida glabrata* under iron-depleted conditions but not under nitrogen starvation. This observation may provide some clues to elucidate the physiological roles of mitophagy in eukaryotes.

Keywords Atg32, Autophagy, *Candida*, Iron, Mitochondria, Mitophagy, Pathogenicity, Yeast

Abbreviations

DHFR	Dihydrofolate reductase
ECL	Enhanced chemiluminescence
GFP	Green fluorescent protein
OD	Optical density
PMSF	Phenylmethane sulfonyl fluoride
ROS	Reactive oxygen species
SD	Synthetic glucose medium
TTBS	Tris-buffered saline containing Tween 20
WT	Wild-type
YPD	Yeast extract peptone dextrose

1 Introduction

Mitochondria mainly contribute to ATP synthesis, and play important roles in cellular energy metabolism. They inevitably consume large amounts of molecular oxygen, and the mitochondrial respiratory chain produces cytotoxic reactive oxygen species (ROS) [1].

ROS are harmful for any cellular components including mitochondria themselves. Therefore, controlling the quality and quantity of mitochondria is essential for maintaining mitochondrial integrity.

Mitophagy, which is a mitochondria-specific degradation via autophagic machinery, is thought to keep mitochondria functional [2–6]. The molecular mechanisms underlying mitophagy have been intensively studied in *S. cerevisiae*, and the findings of *Autophagy* related (*ATG*) genes, responsible for both general autophagy and mitophagy, helped to identify mitophagy-responsible molecules in other fungi [7, 8]. Among the mitophagy-related ATG proteins, Atg32p, a mitochondrial outer membrane protein, is thought to be the key molecule for the initiation of mitophagy [2, 5, 6, 9]. Mitophagy in *S. cerevisiae* cells is induced when these cells are grown under nitrogen-starved conditions, whereas it is more prominent under similar growth conditions in *Pichia pastoris* [7]. In the filamentous fungi *Magnaporthe oryzae*, mitophagy is induced during early stage of conidiation and is required for asexual differentiation [8]. This indicates that mitophagy plays different physiological roles in each fungus; however, the regulatory mechanisms underlying mitophagy are not completely elucidated.

Here we described the materials and methods for the detection of mitophagy in *Candida glabrata*. The opportunistic pathogen *C. glabrata* is a genetically close relative of *S. cerevisiae*, and it is thought to share similar sets of genes including autophagy-related *ATG* genes. Initially, we constructed some *ATG* gene-knock out mutants which would exhibit mitophagy-deficient phenotypes. However, gene targeting by homologous recombination in *C. glabrata* requires much longer sequences homologous to the targeting site than that required in *S. cerevisiae*. To overcome this limitation, a genetically modified *C. glabrata* strain for efficient gene targeting is introduced in this manuscript [10]. Secondary, mitochondria-targeted dihydrofolate reductase (mtDHFR)-GFP construct was introduced into *C. glabrata*. The artificial GFP-fused protein is useful to monitor mitophagy; the vacuolar degradation of mitochondria can be monitored by the generation of processed GFP (Western blot) and by the observation of vacuolar GFP localization in living cells (microscopy). We found that iron-depleted conditions, which partly mimicked blood stream infection, induced mitophagy in *C. glabrata* and that *ATG32* was also essential for mitophagy in the yeast [11]. Nitrogen starvation, which is an authentic mitophagy-inducing condition in *S. cerevisiae*, did not induce mitophagy in *C. glabrata*. The observations indicate genetic similarity and physiological differences underlying mitophagy regulation between these yeast.

2 Materials

2.1 Growth Media for Yeast and *E. coli*

1. Luria Broth (LB): 1% (w/v) Bacto tryptone, 0.5% (w/v) yeast extract, 1% (w/v) NaCl, with 50 μg/mL ampicillin.
2. YPD: 1% (w/v) yeast extract, 2% (w/v) peptone, 2% (w/v) glucose.
3. SD: 0.67% (w/v) yeast nitrogen base, 2% (w/v) glucose.
4. SD-Fe: 0.69% (w/v) Yeast Nitrogen base without Amino acids, without Copper, and without Iron, 2% (w/v) glucose, 100 μM ferrozine [Sigma-Aldrich, 160601] (*See* **Note 1**).
5. SD-N: 0.17% (w/v) yeast nitrogen base without amino acids, 2% (w/v) glucose; for nitrogen-starvation.
6. Nourseothricin (NSTC): (HKI Jena, Germany; 10 mg/mL in water as stock solution).

2.2 Plasmids and Primers

1. Plasmids.

pHIS906 (needed in Sect. 3.1).

p416GPD-mtDHFR-GFP (needed in Sect. 3.2).

Escherichia coli DH5α [F−, ϕ80, lacZΔM15, Δ (lacZYA-argF) U169, hsdR17(rk − mk+), recA1, endA1, deoR, thi-1, supE44, gyrA96, relA1 λ−] was used for plasmid propagation.

2. Primers (5′-3′).

Three primer pairs (—DF and —DR) listed below were used for amplifying the HIS3 cassettes to disrupt *ATG32*, *ATG1*, and *ATG11*, respectively.

ATG32DF; CAAATATAAACTTAAGTGATTTGAGTTTTTAAAGTGTTACAAATTGGTGGTTCAAGGGCCGCTGATCACG.

ATG32DR; GAGATGAGGTAGAAAATCAAATAGATATTAGAAAGGAATAGGTGGGATTTTTCCTACATCGTGAGGCTGG.

ATG1DF; TAACGGATTTGCTAGTTAGGTCTTAAAAATTAGTACTCGAGATGAGCTCCCAAAAGGGCCGCTGATCACG.

ATG1DR; TATTTTTAGGTTATTGTAAAACAACCAATTAATGCATCCCTTTTGGATGAATCTTACATCGTGAGGCTGG.

ATG11DF; TGAACTTGATTTTTTGAACTAGAGCTG ATCTCTGCCATACCATCGCGCTGACACAGGCCGCT GATCACG.

ATG11DR; GAATATATTTAATAATTTCTGTATCGA GAGTTCATTTAATCAAACTATTGACTTCACATCGTG AGGCTGG.

Three primers listed below and pTET12F (AGAAAAC CAGCCTCACGATG) were used for confirmation of the proper cassette integration.

ATG32CHR; ATCTACCCGCCGTTACTCAG.

ATG1CHR; TGTCATCAAGTGGTCGTAGG.

ATG11CHR; TGGAGTAGGTTTGGCACCAC.

HIS3 up -100–80; TATAAAGCTGCGGATGCCTT.

HIS3 down 100-81; GGTAGTTCATGTAGTTAAGA.

GPD pro 200-180; TTTAGTGGATGCCAGGAATA.

mtDHFR CST F; TGGCGCTGGATCTT GACGGTGGGTCTGTTTCTGTACGGGAGAGCATA CTGGGCAAGAGTTTATCATTATC.

mtDHFR CST R; CGCCTTGAAAGCTGATTCG GAGCGGTGGTGGTCGTTGGTGCCCCGCAGACA GATCCGGCCGCAAATTAAA.

2.3 Reagents for Construction of C. glabrata Strains

1. Yeast strains.

   CBS138, ATCC type strain of *C. glabrata*.

   KUE200, *Δtrp1::Scura3 Δhis3::ScURA3 Δura3 FRT-YKU80* (*See* **Note 2**) [10].

2. SD-His agar plate (for selection of *HIS3*+ clone): 0.67% (w/v) yeast nitrogen base without amino acids, 2% (w/v) glucose, 0.077% (w/v) Complete Supplement Mixture (CSM)-histidine, and 2% (w/v) Bacto agar; dissolve all ingredients in water and autoclave. Store solid agar media at room temperature in dark.

3. SD-Trp agar plate (for selection of *TRP1*+ clone): 0.67% (w/v) yeast nitrogen base without amino acids, 2% (w/v) glucose, 0.074% (w/v) Complete Supplement Mixture (CSM)-tryptophan and 2% (w/v) Bacto agar; dissolve all ingredients in water and autoclave. Store solid agar media at room temperature in dark.

4. Yeast extract medium with peptone and dextrose (YPD): 1% (w/v) yeast extract, 2% (w/v) peptone, 2% (w/v) dextrose. Dissolve all ingredients in water and autoclave. Store liquid media at room temperature.

5. 0.15 M Lithium acetate in TE buffer (pH 8.0).
6. 10 mg/mL salmon sperm DNA (ssDNA) in TE buffer (pH 8.0): boiled before use for 10 min and chilled for 10 min to denature double-strand DNA.
7. 52.5% (w/v) Polyethylene glycol (PEG) 4,000 in TE buffer (pH 8.0).
8. YPD plates containing 100 μg/mL NSTC (YPD + NSTC) (HKI Jena, Germany; 10 mg/mL in water as stock solution).
9. 5-FOA plate: Synthetic dextrose medium (0.67% [w/v] yeast nitrogen base [Difco, Becton, Dickinson and Company, 291940], 2% [w/v] glucose, and 2% [w/v] agar, pH 7.0) containing 0.2% (w/v) 5-fluoroorotic acid hydrate (add powder after autoclave, and mix by magnetic stirrer).
10. Sample buffer: 150 mM Tris–HCl, pH 8.8, 6% (w/v) SDS [Wako Pure Chemical Industries, Ltd., 191-07145], 25% (w/v) glycerol, 6 mM EDTA, 0.5% (w/v) 2-mercaptoethanol, 0.05% (w/v) bromophenol blue.

2.4 Reagents for Western Blot

1. Sample buffer: 150 mM Tris–HCl (pH 8.8), 6% (w/v) SDS, 25% (v/v) glycerol, 6 mM EDTA, 0.05% (w/v) bromophenol blue, 0.5% (w/v) 2-mercaptoethanol (add before use).
2. Polyacrylamide gel: SuperSep™ Ace, 12.5% (199-14971, Wako Pure Chemical Industries, Ltd.).
3. Transfer buffer: EzFastBlot (AE-1465, ATTO).
4. Polyvinylidene difluoride (PVDF) membrane: Immun-Blot® PVDF membrane (162-0177, Bio-Rad).
5. Blocking reagent: PVDF Blocking Reagent for Can Get Signal (NYPBR01, TOYOBO).
6. Tris-buffered saline containing Tween 20 (TTBS): 50 mM Tris–HCl, pH 7.6, 0.9% NaCl, 0.1% Tween 20 (P9416, Sigma-Aldrich).
7. Anti-GFP antibody: Living Colors® A.v. Monoclonal Antibody (JL-8) (632380, Clontech).
8. Anti-Pgk1 antibody: Rabbit IgG against baker's yeast 3-Phosphoglyceric phosphokinase (NE130-7S, Nordic Immunology). The antibody was raised against *S. cerevisiae* Pgk1. It cross-reacted to *C. glabrata* Pgk1.
9. First antibody dilution buffer: Can Get Signal® Immunoreaction Enhancer Solution 1 (NKB-201, TOYOBO).
10. HRP-conjugated anti-mouse IgG antibody: Goat anti-Mouse IgG (H + L) Secondary Antibody, HRP (32430, Thermo Fisher Scientific).

11. HRP-conjugated anti-rabbit IgG antibody: Stabilized Goat Anti-Rabbit HRP conjugated (1858415, PIERCE).
12. Second antibody dilution buffer: Can Get Signal® Immunoreaction Enhancer Solution 2 (NKB-301, TOYOBO).
13. ECL Western Blotting Detection Kit: ImmunoStar LD (296-69901, Wako Pure Chemical Industries, Ltd.).

3 Methods

3.1 Introduction of HIS3 Gene into the KUE200 Strain

The DNA fragment used to replace the *ATG32*, *ATG1*, or *ATG11* ORF with *HIS3* was amplified from the plasmid pHIS906 using the primer sets ATG32DF and ATG32DR, ATG1DF and ATG1DR, or ATG11DF and ATG11DR, respectively (Fig. 1a). Each amplified fragment was used to transform the KUE200 strain [10].

DNA polymerase for PCR: Phusion High-Fidelity DNA Polymerase (M0530S, New England Biolabs).

Thermal cycling: 95 °C 3 min, 35 × (95 °C 1 min, 55 °C 15 s, 72 °C 1 min), 72 °C 3 min, 4 °C forever.

1. Streak the KUE200 cells on YPD + NSTC agar plate, and grow for 2–3 days at 37 °C.
2. Pick a few colonies, resuspend them in 1 mL of YPD + NSTC liquid medium, and culture overnight with shaking at 37 °C.
3. Inoculated the overnight culture in 20 mL of fresh YPD + NSTC (starting optical density at 600 nm [OD_{600}] of 0.4), and shake the cell suspension to an OD_{600} of 1.0, approximately for 2–3 h at 37 °C.
4. Harvest the cells by centrifugation at 6,000 × *g* for 5 min. Rinse the cell pellet with 10 mL of TE buffer, and harvest by centrifugation. Resuspend the cells in 10 mL of 0.15 M lithium acetate dissolved in TE buffer (LiOAc/TE), and shake the tube slowly for 1 h at 30 °C.
5. Harvest the cells by centrifugation and resuspend the cells in 400 μL of 0.15 M LiOAc. Dispense 60 μL of the suspension to a new 1.5 mL tube, supplemented with 5–10 μg (3 μL) of the disruption cassettes and 20 μg (2 μL) of carrier DNA (salmon sperm DNA [Wako], boiled before use for 10 min and chilled for 10 min to denature double-strand DNA), and mix gently.
6. Incubate the cell suspension for 30 min at 37 °C.
7. Add 120 μL of 52.5% (w/v) polyethylene glycol 4,000 (solubilized in 0.15 M LiOAc), and mix thoroughly by pipetting. Incubate the cell suspension for 45 min at 37 °C.
8. After mixing carefully, heat-shock the cell suspension by incubating for 45 min at 42 °C and then spread onto SD-His agar plates, and incubate the plate at 37 °C for a few days.

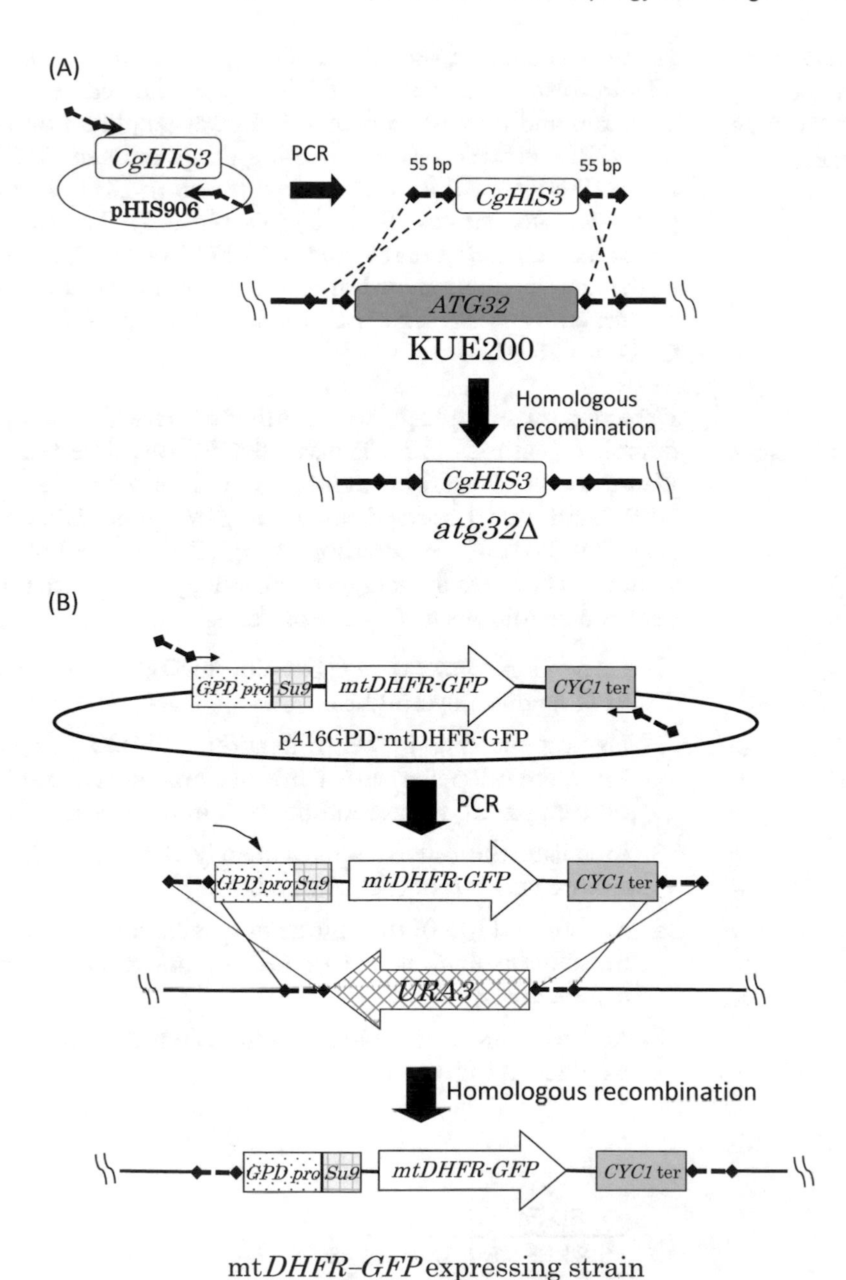

Fig. 1 Schematic illustration of the genetic modification: (**a**) disruption of *ATG32,* (**b**) introduction of mtDHFR-GFP

Disruption of the *ATG* genes was confirmed by PCR using primers pTET12F and ATG32CHR for *ATG32*, pTET12F and ATG1CHR for *ATG1,* and pTET12F and ATG11CHR for *ATG11*, respectively.

3.2 Introduction of mtDHFR-GFP into KUE200 and ATG Gene-Deleted Mutants

A transformation cassette containing the mitochondria-targeted dihydrofolate reductase (DHFR)-GFP flanked by sequences upstream and downstream of *URA3* was amplified by PCR from p416GPD-mtDHFR-GFP [6] using the primers mtDHFR CST F and mtDHFR CST R, and used to transform KUE200 and *ATG* gene-disrupted strains (Fig. 1b) (*See* **Note 3**). The transformants which lack *URA3* were selected on 5-FOA plate. Correct insertion of the amplified fragment into the correct chromosomal locus was confirmed by PCR using the primers HIS3 up −100 to −80 and GPD pro 200-180.

3.3 Detection of Mitophagy by Western Blotting

Detection of mitophagy was conducted according to previously described methods [12]. Processed GFP was detected in WT *C. glabrata* cells grown iron-depleted conditions for ~48 h, while no GFP band was observed in all *atg32*Δ gene-deleted mutants (Fig. 2) (no GFP processing in *atg1*Δ and *atg11*Δ, data not shown). These results suggest mitophagy induction under iron-depleted conditions in *C. glabrata* cells.

1. Grow *C. glabrata* cells expressing mtDHFR-GFP in 1 mL of YPD medium overnight.
2. Harvest the cells by centrifugation at 6,000 × *g* for 5 min. Rinse the cell pellet with 1 mL of sterile water, and harvest by centrifugation. Resuspend the cells in the test medium.
3. Inoculate the cells at approximately 3 × 10^5 cells/mL, and culture in 5 mL of the test medium.
4. Measure OD_{600} of the culture at specific time points. Dispense the aliquots equivalent to 1 OD_{600} unit in 1.5 mL microcentrifuge tubes.
5. Add trichloroacetic acid (10% final concentration), and incubate for 10 min on ice.

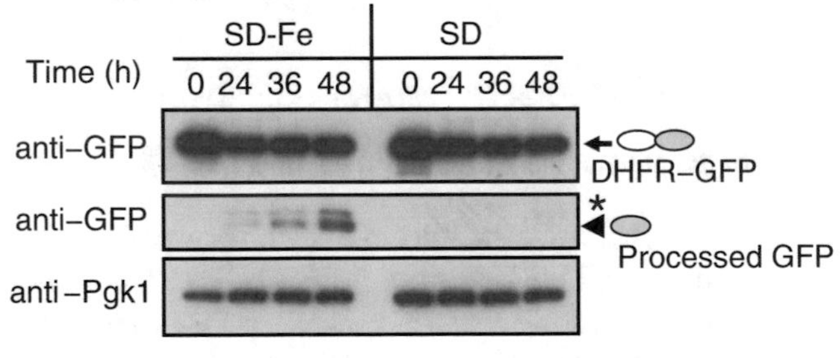

Fig. 2 KUE200 (wild-type) cells expressing mtDHFR-GFP were grown in SD or SD-Fe medium, and collected at the indicated time points, and then subjected to Western blot analysis with anti-GFP antibody. mtDHFR-GFP and the processed GFP moiety are indicated by the *arrow* and *arrowhead*, respectively. The generation of processed GFP indicates the vacuolar degradation of mtDHFR-GFP. Pgk1 was monitored as a sample loading control. Nonspecific bands are designated by an *asterisk*

6. Harvest the cells by centrifugation at 21,000 × *g* for 10 min. After washing the pellet fraction twice with 1 mL of ice-cold acetone, place the tube open for air dry.
7. Resuspend the air-dried cell pellet in 50 μL of sample buffer, and disrupt the cells by vortexing with an equal volume of acid-washed glass beads for 3 min.
8. After denaturation of the homogenate at 100 °C for 3 min, load 3 μL aliquots of the sample onto a 12.5% polyacrylamide gel, and perform electrophoresis.
9. Transfer the protein sample to PVDF membrane by semi-dry western blot transfer procedure (15 V, 60 min for 8.5 × 6 cm size membrane).
10. After quick rinse of the membrane in TTBS buffer, incubate the PVDF membrane with the blocking reagent for 1 h at room temperature.
11. Perform a first antibody incubation with anti-GFP antibody (1:25,000 dilution) by incubating for 1 h at room temperature. [For loading control; anti-Pgk1 antibody (1:100,000 dilution).]
12. Wash the membranes three times each for 10 min in TTBS.
13. Perform a secondary incubation with HRP-conjugated anti-mouse IgG (1:20,000 dilution; Thermo Fisher Scientific, 32430) for 1 h at room temperature. [For loading control; HRP-conjugated anti-rabbit antibody (1:20,000 dilution).]
14. Wash the membrane three times each for 10 min in TTBS.
15. Mix the solution A and B (1:1) of the ECL Western Blotting Detection Kit, and incubate the membrane in the mixed solutions for 2 min.
16. Perform imaging analysis of the membrane using a C-DiGit Blot Scanner and ImageStudio software (LI-COR Biosciences, Lincoln, NE). MtDHFR-GFP and processed GFP were detected as bands that migrated at molecular masses of approximately 50 and 28 kDa, respectively.

3.4 Detection of Vacuolar Degradation of Mitochondria by Fluorescent Microscopy

mtDHFR-GFP-expressing WT and *atg32*Δ cells were cultured in SD-Fe medium for 2 days, then the GFP localization was observed by fluorescence microscopy (IX-81, Olympus). The vacuolar localization of GFP was observed with WT cells but not with *atg32*Δ mutant cells (Fig. 3). The observation suggests that mitophagy was induced under iron-depleted conditions in *C. glabrata* cells and that *ATG32* is responsible for mitophagy in *C. glabrata* as well as in *S. cerevisiae*.

1. Grow the *C. glabrata* cells in 1 mL YPD medium overnight.

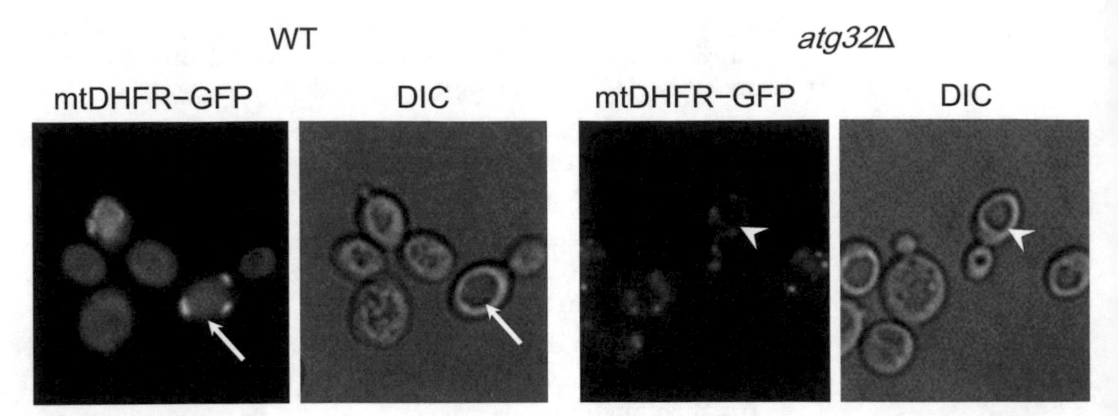

Fig. 3 Fluorescence microscopy observation of GFP. mtDHFR-GFP-expressing WT and *atg32Δ* cells were cultured in SD-Fe medium for 2 days, then the GFP localization was observed by fluorescence microscopy. The cells were incubated for 4 h in the presence of 1 mM PMSF prior to the observation. DIC shows differential interference contrast of the identical field to the left panel. Diffused GFP inside vacuole (*white arrows*) indicates mitochondrial degradation in WT cells, while almost complete loss of vacuolar GFP signal (*white arrow heads*) indicates lack of mitochondrial degradation in *atg32Δ* cells

2. Harvest the cells by centrifugation at 6,000 × *g* for 5 min. Resuspend the cell pellet with 1 mL of PBS, and harvest again by centrifugation. Resuspend the cell pellet again in 5 mL of SD-Fe, and shake at maximum speed (180 rpm) for 2 days at 37 °C.
3. Put 5 μL of 1 M PMSF (final concentration; 1 mM), and incubate for 4 h at 37 °C prior to microscopic observation.
4. Harvest the cells by centrifugation at 6,000 × *g* for 5 min. Resuspend the cells in 20 μL of mounting solution (VECTASHIELD H-1000, Vector Laboratories catalog #H-1000). Confirm the localization of vacuole in DIC image, and the colocalization of GFP signal by the filter set for GFP (1,000× image) (IX-81, Olympus).

4 Notes

1. Ferrozine, a chelator of both ferric and ferrous ion, was used for iron-depletion. It was impossible to make iron-depleted yeast cells without a cherating compound. Intensive cell washes before medium changes were not sufficient to prevent iron carryover from the pre-culture medium. Moreover, intracellular stored iron could not be immediately exhausted.
2. *YKU80* gene is thought to suppress homologous recombination in *C. glabrata* [10]. The KUE200 strain harbors a *SAT1* flipper cassette between the first and second ATG codon of *YKU80*. The *SAT1* flipper cassette integration knocked down the expression of *YKU80* and resulted in an enhanced efficiency of homologous recombination. The homologous recombination efficacy

in KUE200 was 13% (the number of correct integrations per total transformants) fusing a ~60 bp homologous flanking DNA at both the 5′- and 3′-end of the transformation cassettes. In ACG4, a strain with identical genetic background but intact *YKU80*, this length of homologous DNA provided no correct integration.

To prevent pop-out of the *SAT1* flipper cassette during transformation, the KUE200 strain must be routinely kept in the presence of nourseothricin (NSTC). The function of *YKU80* is thought to be restored after pop-out of the *SAT1* flipper cassette by incubation in NSTC-free medium.

3. The plasmid, p416GPD-mtDHFR-GFP self-replicates in *S. cerevisiae*, and enables mtDHFR-GFP expression in the yeast [6]. The plasmid was temporally maintained in *C. glabrata* but did not produce detectable mtDHFR-GFP. Therefore, the mtDHFR-GFP cassette was integrated to chromosome to be maintained in *C. glabrata* cells. KUE200 is auxotrophic for both histidine and tryptophan (*his3*Δ and *trp1*Δ). *HIS3* and *TRP1* were already used as selectable markers to disrupt and reintroduce the *ATG* genes, respectively (the reintroduction of *ATG32* was not shown). *URA3* was introduced into the *HIS3* locus, and it remained at the locus in KUE200. The mtDHFR-GFP cassette was designed to disrupt *URA3*. The resulting *ura3*-clone could be selected on a 5-fluoroorothidine acetate (FOA) plate on which *URA3*+ clones are unable to grow.

Acknowledgement

We thank Koji and Noriko Okamoto (Graduate School of Frontier Biosciences, Osaka University, Suita, Osaka 565-0871, Japan) for providing yeast strains and plasmids. This research was supported by JSPS KAKENHI Grant Numbers 26790428 (KT), 24590540 (HN), and 26860296 (MN). This work was partly supported by a grant from the Ministry of Health, Labour and Welfare of Japan (H26-shinkoujitsuyouka-ippan-010). The authors would like to thank Enago (www.enago.jp) for the English language review.

References

1. Turrens JF (2003) Mitochondrial formation of reactive oxygen species. J Physiol 552:335–344
2. Kurihara Y, Kanki T, Aoki Y, Hirota Y, Saigusa T, Uchiumi T, Kang D (2012) Mitophagy plays an essential role in reducing mitochondrial production of reactive oxygen species and mutation of mitochondrial DNA by maintaining mitochondrial quantity and quality in yeast. J Biol Chem 287:3265–3272
3. Narendra DP, Jin SM, Tanaka A, Suen D-F, Gautier CA, Shen J, Cookson MR, Youle RJ (2010) PINK1 is selectively stabilized on impaired mitochondria to activate Parkin. PLoS Biol 8:e1000298

4. Narendra D, Tanaka A, Suen D-F, Youle RJ (2008) Parkin is recruited selectively to impaired mitochondria and promotes their autophagy. J Cell Biol 183:795–803
5. Kanki T, Wang K, Cao Y, Baba M, Klionsky DJ (2009) Atg32 is a mitochondrial protein that confers selectivity during mitophagy. Dev Cell 17:98–109
6. Okamoto K, Kondo-Okamoto N, Ohsumi Y (2009) Mitochondria-anchored receptor Atg32 mediates degradation of mitochondria via selective autophagy. Dev Cell 17:87–97
7. Aihara M, Jin X, Kurihara Y, Yoshida Y, Matsushima Y, Oku M et al (2014) Tor and the Sin3-Rpd3 complex regulate expression of the mitophagy receptor protein Atg32 in yeast. J Cell Sci 127:3184–3196
8. He Y, Deng YZ, Naqvi NI (2013) Atg24-assisted mitophagy in the foot cells is necessary for proper asexual differentiation in *Magnaporthe oryzae*. Autophagy 9:1818–1827
9. Kanki T, Wang K, Baba M, Bartholomew CR, Lynch-Day MA, Du Z et al (2009) A genomic screen for yeast mutants defective in selective mitochondria autophagy. Mol Biol Cell 20:4730–4738
10. Ueno K, Uno J, Nakayama H, Sasamoto K, Mikami Y, Chibana H (2007) Development of a highly efficient gene targeting system induced by transient repression of YKU80 expression in Candida glabrata. Eukaryot Cell 6:1239–1247
11. Nagi M, Tanabe K, Nakayama H, Ueno K, Yamagoe S, Umeyama T, Ohno H, Miyazaki Y (2016) Iron-depletion promotes mitophagy to maintain mitochondrial integrity in pathogenic yeast *Candida glabrata*. Autophagy 12:1259–1271
12. Kanki T, Kang D, Klionsky DJ (2009) Monitoring mitophagy in yeast: the Om45-GFP processing assay. Autophagy 5:1186–1189

Methods in Molecular Biology (2018) 1759: 173–175
DOI 10.1007/7651_2018_156

Published online: 27 May 2018

Dopaminergic Neuron-Specific Autophagy-Deficient Mice

Shigeto Sato and Nobutaka Hattori

Abstract

None of the current genetic Parkinson's disease (PD) models in mouse recapitulates all features of PD. Additionally, only a few of these models develop mild dopamine (DA) neurodegeneration. And the most parsimonious explanation for the lack of DA neurodegeneration in genetic PD models is a compensatory mechanism that results from adaptive changes during development, making it hard to observe the degenerative phenotype over the life span of mice. Here, we characterize DA neuron-specific autophagy-deficient mice and provide in vivo evidence for Lewy body formation. Atg7-deficient mice demonstrate typical Lewy pathology, including endogenous synuclein and neuronal loss, which resembles PD. Furthermore DA levels are affected by dopaminergic neuronal loss. The age-related motor dysfunction and pathology in DA neurons suggest that impairment of autophagy is a potential mechanism underlying the pathology of PD.

Keywords Atg7, Autophagy, Dopaminergic neuron, Lewy body, Mouse model, Parkinson's disease

1 Introduction

Macroautophagy is a highly conserved bulk protein degradation pathway in eukaryotes. Cytoplasmic proteins and organelles are engulfed within autophagosomes, which fuse with the lysosome, where they are degraded along with their cargo. For example, several lines of evidence indicate that synuclein is predominantly degraded by autophagy and mutant forms of synuclein are dependent on the autophagy-lysosome pathway for their clearance [1–3]. Although the phenotypes of mice harboring brain-specific deletion of *Atg5* or *Atg7* reveal the critical role of autophagy in the removal of aggregated proteins [4, 5], Friedman et al. [6] demonstrated that dopaminergic neuron-specific autophagy deficiency leads to the restrictive presynaptic accumulation of synuclein in the dorsal striatum, suggesting that impaired autophagy plays a role in Parkinson's pathogenesis. In dopaminergic neurons, the primary site of endogenous pathology, no detailed reports to date have examined the endogenous synuclein accumulation in dopaminergic cell bodies and neurites, which is associated with Lewy pathology. To understand the effects of autophagy impairment on dopaminergic neurons, we characterized conditional knockout mice harboring a tyrosine hydroxylase (TH) neuron-specific deletion of *Atg7* and observed their age-related pathological and motor phenotypes [7].

2 Materials and Methods

2.1 Animals

All animals were kept in a pathogen- and odor-free environment, which was maintained under a 12 h light/dark cycle at ambient temperature. Floxed Atg7 mice were characterized previously and were crossed with TH-Cre mice carrying the knock-in construction containing TH fused to Cre in the 3′end to generate $Atg7^{flox/flox}$:TH-Cre mice.

2.2 Behavioral Tests

Locomotor behavior was assessed in mice from 90 to 120 weeks of age. Accelerating rotarod tests were performed on a rotarod machine with automatic timers and falling sensors (MK-660D, Muromachi Kikai). $Atg7^{flox/flox}$ mice and $Atg7^{flox/flox}$:TH-Cre mice were placed on a 3 cm diameter rotating rod covered with rubber, and rotation was accelerated from 3 to 35 rpm over 5 min. Fall latency was recorded, and the first fall latency of the third trial was used for analysis. The runway test was performed using a narrow, horizontally fixed beam (1 cm wide, 80 cm long, held at a height of 40 cm from the table).

2.3 Histological Analyses

Mice were perfused with 4% paraformaldehyde, and their brains were immersion-fixed at 4 °C for 36 h. The fixed samples were cryoprotected with 20% sucrose and sliced on a freezing microtome to obtain 40-μm-thick floating sections. For double immunohistochemistry of p62 and TH, sections were initially incubated with the anti-p62 antibody (PROGEN Biotechnik, GmbH) and visualized with diaminobenzidine (DAB)-containing nickel ammonium sulfate (DABNi), which generates dark purple precipitates. The sections were observed on a VS120 (Olympus, Tokyo, Japan). For double immunofluorescence of p62 (green) and TH (red), floating sections were incubated with rabbit anti-TH antibody (657,012, Calbiochem, Germany) and guinea pig anti-p62 antibody (PROGEN Biotechnik, GmbH). They were then incubated with anti-guinea pig IgG conjugated with Alexa Fluor 488 and anti-rabbit IgG conjugated with Alexa Fluor 546. Fluorescent signals were captured by LSM 780 confocal microscopy (Zeiss).

2.4 HPLC Analysis

Dorsal striata from mice brain were dissected, quickly frozen on dry ice, and then homogenized with 0.5 mL of 0.2 M perchloric acid containing 100 μM EDTA-2Na per 100 mg wet tissue. Samples were centrifuged at 20,000 × *g* for 15 min at 4 °C. The supernatant was collected and analyzed by HPLC.

2.5 Phenotype and Pathology of Dopaminergic Neuron-Specific Autophagy-Deficient Mice

- In Atg7$^{flox/flox}$:TH-Cre midbrains, the quantification of Atg7 after normalization by actin showed about 14% residual amount, compared to controls.
- Atg7$^{flox/flox}$:TH-Cre mice began to show impairment in motor coordination tasks around 100 weeks and apparent motor behavioral deficits around 110 weeks. These clinical abnormalities could be demonstrated by the runway test and rotarod test.
- In contrast to Atg7$^{flox/flox}$ mice, which exhibited well-coordinated movement and almost no slips of the forepaw or hindpaw from the beam, the Atg7$^{flox/flox}$:TH-Cre mice could hardly move on the beam and slipped frequently.
- In the accelerating rotarod test, the fall latency was reduced in Atg7$^{flox/flox}$:TH-Cre mice.
- In Atg7$^{flox/flox}$:TH-Cre mice, these neurons contained eosinophilic aggregates, which are characteristic of Lewy body including ubiquitin and p62.
- The number and size of these inclusions were gradually increased in neurites rather than soma with aging.
- Synuclein deposition is preceded by p62 and resulted in the formation of inclusions containing synuclein and p62.
- Endogenous synuclein colocalizes with p62-positive inclusions in Atg7$^{flox/flox}$:TH-Cre mice.
- About 90% synuclein inclusions are located in TH fibers.
- The reduction in TH cell number was most prominent in the center area of substantia nigra pars compacta.
- High-performance liquid chromatography (HPLC) revealed a reduction in striatal dopamine levels and metabolites in Atg7$^{flox/flox}$:TH-Cre versus control mice.

References

1. Webb JL, Ravikumar B, Atkins J, Skepper JN, Rubinsztein DC (2003) Alpha-synuclein is degraded by both autophagy and the proteasome. J Biol Chem 278:25009–25013
2. Spencer B et al (2009) Beclin 1 gene transfer activates autophagy and ameliorates the neurodegenerative pathology in alpha-synuclein models of Parkinson's and Lewy body diseases. J Neurosci 29:13578–13588
3. Yu WH et al (2009) Metabolic activity determines efficacy of macroautophagic clearance of pathological oligomeric alpha-synuclein. Am J Pathol 175:736–747
4. Hara T et al (2006) Suppression of basal autophagy in neural cells causes neurodegenerative disease in mice. Nature 441:885–889
5. Komatsu M et al (2006) Loss of autophagy in the central nervous system causes neurodegeneration in mice. Nature 441:880–884
6. Friedman LG et al (2012) Disrupted autophagy leads to dopaminergic axon and dendrite degeneration and promotes presynaptic accumulation of alpha-synuclein and LRRK2 in the brain. J Neurosci 32:7585–7593
7. Sato S et al (2018) Loss of autophagy in dopaminergic neurons causes Lewy pathology and motor dysfunction in aged mice. Sci Rep 8:2813

INDEX

Nobutaka Hattori and Shinji Saiki (eds.), *Mitophagy: Methods and Protocols*, Methods in Molecular Biology, vol. 1759,
https://doi.org/10.1007/978-1-4939-7750-5, © Springer Science+Business Media, LLC, part of Springer Nature 2018

N

O

P

R

Printed in the United States
By Bookmasters